A comprehensive review of methods for the channel allocation problem

Jayrani Cheeneebash and Harry Coomar Shumsher Rughooputh

Published in 2014 by African Minds
4 Eccleston Place, Somerset West, 7130, Cape Town, South Africa
info@africanminds.org.za
www.africanminds.org.za

ISBN: 978-1-920677-53-4

ORDERS:
African Minds
info@africanminds.org.za
www.africanminds.org.za

For orders from outside Africa:
African Books Collective
PO Box 721, Oxford OX1 9EN, UK
orders@africanbookscollective.com

To my son Devesh who is now an angel.

The soul is neither born nor does it ever die;
having come into being once, it never ceases to be.
It is unborn, eternal, permanent, and primeval.
(The Bhagavad Gita)

Table of contents

List of tables vii

List of figures viii

Abbreviations ix

Preface x

Part I The channel allocation problem 1

1 Introduction 3

2 The channel allocation problem 11
- 2.1 Channel allocation . . . 11
  - 2.1.1 Interference . . . 11
- 2.2 Formulation of the CAP . . . 13
- 2.3 Channel allocation schemes . . . 13
  - 2.3.1 Fixed channel allocation . . . 14
  - 2.3.2 Dynamic channel allocation . . . 14
  - 2.3.3 Comparison between FCA and DCA . . . 15
  - 2.3.4 Hybrid channel allocation . . . 16

3 Mathematical formulation of the CAP 19
- 3.1 Introduction . . . 19
  - 3.1.1 Problem formulation – FCA . . . 19
  - 3.1.2 Problem formulation – DCA . . . 20
  - 3.1.3 Problem formulation – HCA . . . 20
  - 3.1.4 An example of a four-cellular network . . . 20
- 3.2 Lower bounds for the CAP . . . 22
- 3.3 Summary . . . 22

Part II Methods used to solve channel allocation problem 27

4 Graph theory 29
- 4.1 Graph colouring approach . . . 29
- 4.2 Summary . . . 32

5 Neural networks 35
- 5.1 Neural network approach . . . 35
- 5.2 Summary . . . 39

6 Heuristics 43
- 6.1 Heuristic techniques . . . 43

6.1.1 Summary 45
6.2 Simulated annealing 46
6.2.1 Summary 47
6.3 Tabu search 47
6.3.1 Summary 48

7 Evolutionary methods **51**
7.1 Genetic algorithm 51
7.1.1 Summary 53
7.2 Evolutionary strategy 53
7.2.1 Summary 54

8 Q-learning and other algorithms **59**
8.1 Q-learning approach 59
8.1.1 Summary 60
8.2 Other algorithms 60

9 The channel allocation problem from a multi-objective view **63**
9.1 Multi-objective approach 63
9.2 Experimental results 63
9.3 Comparison of different approaches 65
9.4 Summary 65

List of tables

2.1 Comparison between FCA and DCA 16

3.1 Channel demand 21
3.2 Constraints based on compatibility matrix C 21
3.3 Call list 21
3.4 Frequency at every cell 21
3.5 Frequency channel separation constraint C, where $k \geq 1$, $m \leq 3$ 22
3.6 Demand vectors D_1 and D_2 23
3.7 Interference constraints 23

9.1 Comparison of results for the different approaches to solve the CAP 65

List of figures

1.1 Clustered cells in a cellular structure 5
1.2 A one-dimensional cellular array 6

2.1 Three cellular structure 13
2.2 Structure of the allocation matrix A 14

3.1 Structure of cellular systems 22

4.1 Example of a four-cell graphical representation 29

9.1 Pareto front for problem 2 (Table 3.7) 64

Abbreviations

a.c.c adjacent channel interference

BCA borrowing channel assignment

BS base station

BSC base station controller

CAP channel allocation problem

c.c.c co-channel interference

CD code division

CSA chaotic simulated annealing

c.s.c co-site interference

DCA dynamic channel allocation

FCA fixed channel allocation

FD frequency division

FEA frequency exhaustive assignment

FM frequency modulation

HCA hybrid channel allocation

ISDN Integrated Series Digital Network

MSC mobile switching centre

MTSO mobile telephone switching office

PSTN public switched telephone network

REA requirement exhaustive assignment

RSD randomised saturation degree

SCSA stochastic chaotic simulated annealing

SSA stochastic simulated annealing

TD time division

Preface

The study of the channel allocation problem has been successfully developed during the last decade. Several techniques such as genetic algorithm, artificial neural network, simulated annealing, tabu search and others have been used.

This book is devoted to compiling all the techniques that have been used to solve the channel allocation problem. Each of the methods is described fully in a manner that explains the essential parts of how the techniques are formulated and applied in solving the problem. This textbook will be helpful to students studying communications or researchers as it compiles all the techniques used since this problem was first solved until the present.

The book has nine chapters and is divided into two parts.

Part I describes the channel allocation problem and the different allocation schemes. It gives the reader basic information about channel allocation and how it is formulated. The second chapter gives the mathematical formulation of each channel allocation scheme and how it is described as an optimisation problem. The theoretical lower bounds are also given, as these provide a guide to comparing the different methods.

Part II brings together several techniques that have been used in the last decade. Most of the algorithms have been applied to fixed channel allocation schemes as well as to some cases involving dynamic allocation. In all cases of fixed allocation schemes, the Philadelphia problem has been used as a benchmark. This part ends with a chapter which compares the results obtained by the different techniques and thus gives the reader an idea of the performance of each method.

This book can be useful to researchers in the field of wireless or mobile communication. It can also be useful to communication engineers who, in their daily lives, face real-world optimisation problems. They can use this book as guide as it reviews most methods. Moreover, students in communication engineering will find this book interesting.

Reduit, Mauritius
June 2014

Jayrani Cheeneebash
Harry Coomar Shumsher Rughooputh

Part I
The channel allocation problem

In this part, we give a theoretical background on the channel allocation problem and describe the cellular concept and the different interference that occurs between cells. We also describe the different channel allocation schemes, namely fixed channel allocation, dynamic channel allocation and hybrid channel allocation. A comparison between fixed channel allocation and dynamic channel allocation is given. The last chapter of this part gives a mathematical formulation of the problem and how it is considered as an NP-hard optimisation problem. Also, the theoretical lower bounds for the some benchmark problems are given so that later in the following part of the book, the efficiency of the different algorithms used can be assessed.

Chapter 1

Introduction

During the past decades, we have been witnessing the revolution in telecommunications devices and their impact on our daily life. Indeed, mobile computing has emerged as an important topic of research. People now communicate via an increasing number of devices, many of which are mobile, and the number of mobile users keeps on increasing worldwide. To be able to satisfy users, wireless networks have been used to provide integrated services, but also to support the facilities of dynamically locating the mobile terminals and enabling efficient message routing among them. Therefore, an efficient allocation of channels for proper communication is important due to bandwidth limitation.

Modern wireless networks are organised in geographical cells, each controlled by a base station (BS). The use of cells can increase the capacity of a wireless network, allowing more users to communicate simultaneously. The number of simultaneous calls a mobile wireless system can accommodate is essentially determined by the total spectral allocation for that system and the bandwidth required for transmitting signals used in handling a call.

MacDonald (1979) introduced the cellular concept in which the radio coverage area of a base station is represented by a cell. The cellular networks involve a relatively simple architecture within which most of the communication aspects of wireless systems can be studied. A cellular network consists of a large number of wireless subscribers who have cellular phones that can be used in cars, in buildings, on the streets and almost anywhere. There is also a number of fixed base stations, arranged to provide coverage via wireless electromagnetic transmission of the cell phones.

A regular hexagon is chosen to represent a cell because it covers a larger area with the same centre-to-vertex distance compared to a square or equilateral triangle. Consequently, fewer hexagonal cells need to be placed in a cellular structure to cover a given geographical area and these cells are grouped into clusters. The entire block of frequencies is completely allocated to each cluster and the cells in each cluster use different frequencies. In this way, the frequency in the bandwidth is reused.

Two major fields of interest in cellular mobile networks have evolved, namely mobility management and bandwidth management. Mobility management consists of two basic components: location management and handoff management. Location management handles the tracking of mobile terminals and the channelling of incoming calls to the mobiles. Handoff management deals with providing continuity of a call with the required quality of service, even when the users move from the coverage area of one base station to that of another base station.

With the ever-increasing number of mobile users and a pre-assigned communication bandwidth, the problem of efficiently using the radio spectrum for cellular mobile communication has become a critical research issue (Chen et al., 2002, 2003; Sarkar and Sivaranjan, 1998). Channel interference in the reuse of radio spectrum cells is the major factor which needs to be considered while solving the channel allocation problem (CAP). Neglecting other influencing factors, we assume that the channel interference is primarily a function of frequency and distance. A channel can simultaneously be used by multiple base stations if their mutual separation is more than the reuse distance, which is the minimum distance at which two signals of the same frequency do not interfere. In a cellular environment, the reuse distance is usually expressed in units of number of cells. There are three types of interference that exist in a cellular environment, as described in Chapter 2. The task of assigning frequency channels to the cells that satisfies the frequency separation constraints with a view to avoiding channel interference and using as little bandwidth as possible is known as the *channel allocation problem* (CAP). In its most general form, the CAP is equivalent to the generalised graph-colouring problem (Metzger, 1970) which is a well-known NP-complete problem (Hale, 1980). Because of the nature of the CAP, much research has been carried out in order to develop time-efficient heuristic or approximate algorithms, which cannot guarantee optimal solutions, however research in this field dates from the early 1970s to the present. (Gamst and Rave, 1982) defined the general form of the channel allocation problem in an arbitrary inhomogeneous cellular radio network as an optimisation problem. Several methods have been used to solve the CAP problem. These include:

- Graph colouring Method (Gamst and Rave, 1982; Sivaranjan et al., 1989; Yeung, 2000).
- Neural network Approach (Kunz, 1991; Funabiki and Takefuji, 1992; Kim et al., 1997; Lázaro and Girma, 2001).
- Genetic algorithm (Kim et al., 1996; Lai, 1996; Lima et al., 2002; Ghosh et al., 2003; Yen and Hanzo, 2004*b*).
- Evolutionary strategy (Creput et al., 2005; Sandalidis et al., 1998; Vidyarthi et al., 2005).
- Local search algorithm (Wang and Rushforth, 1996; Kendall and Mohamad, 2004*c*).
- Simulated annealing (Li and Wang, 2001; Wang and Rushforth, 1996; Santos et al., 2001).
- Tabu search (Castelino et al., 1996; Hao and Perrier, 1996; Hao et al., 1998).
- Q-learning approach (Nie and Haykin, 1999*a*,*b*).

Among other approaches used, (Ghosh et al., 2006) deal with a coalesced channel allocation problem.

To compare the results of the different methods used, authors have solved the Philadelphia problem, also known as the 21-cells problem. From the literature it can be seen that some methods do not always give the optimal solution for certain cases whereas others have a slow rate of convergence or need expertise to be applied (Beckmann and Killat, 1999).

Various schemes have been devised for solving the CAP. In the fixed channel allocation (FCA) scheme, the geographical area is partitioned into a number of cells, and a number of channels is assigned to each cell according to some reuse pattern depending on the desired signal quality. However, the FCA does not adapt well to changing traffic conditions and changing user distributions. The dynamic channel allocation (DCA) scheme provides more flexibility and traffic adaptability by placing all channels in a pool and assigning channels to new calls as needed. However, DCA is also less efficient under high load conditions due to its higher cost complexity. To overcome the drawback at high load conditions, hybrid channel allocation (HCA) was designed by combining FCA and DCA schemes (Yue, 1991).

The advent of the cellular concept was a major breakthrough in the development of wireless mobile communication. Mobile communication has improved our lives considerably and the most popular device at the moment is the cellular phone. The major advantages of the wireless link over the wired link are:

1. Faster speed of deployment;
2. Accessibility to difficult areas; and

3. Low marginal cost and effort in adding or removing a subscriber compared to the cost required to install cables for wired access.

On the other hand, the main disadvantages are that wireless signals can be received anywhere within the coverage area, and since it is easier to gain unauthorised network access, security may be at risk. Besides, it is more difficult to transmit data at a high rate in a wireless channel than a wired channel since the wireless channel is more hostile.

(MacDonald, 1979) introduced the cellular concept where the radio coverage area of a base station is represented by a cell. During the early part of the evolution of the cellular concept, the system designers recognised the concept of all cells having the same shape to be helpful in systematising the design and layout of the cellular system. In the Bell Laboratories paper (MacDonald, 1979), four possible geometrical shapes were discussed: the circle, the square, the equilateral triangle and the regular hexagon. As mentioned in Chapter 1, we use a regular hexagon to represent a cell. These cells are placed in a cellular structure covering a geographical area as shown in Figure 1.1.

The cells in a geographical area are grouped into clusters. The entire block of frequencies is completely allocated to each cluster and the cells in each cluster use different frequencies. In this way, the limited block of frequency spectrum is reused. The concept of frequency is illustrated in Figure 1.1, where the cluster size is seven cells and a set of co-channel cells – that is, cells using the same frequency – is shown shaded. The co-channel reuse ratio is given as:

$$\frac{R_u}{R_b} = \sqrt{3N_c}, \tag{1.1}$$

where R_u is the distance between the two closest co-channel cells, R_b is the radius of the cell and N_c is a positive integer representing the number of cells per cluster (Wong, 2003).

Each cell has a base station and a number of mobile terminals (e.g. for mobile phones, palmtops, laptops or other mobile devices). The base station is equipped with radio transmission and reception equipment. The mobile terminals within a cell communicate through wireless links with the base stations associated with the cell. Several base stations are connected to the base station controller (BSC) via microwave links or dedicated leased lines. The BSC contains logic for radio resource management of the base stations under its control. It is also responsible for transferring an ongoing call from one base station to another as a mobile user moves from cell to cell.

Several BSCs are connected to a mobile switching centre (MSC), also known as mobile telephone switching office (MTSO). The MSC/MTSO is responsible for setting up and terminating calls to and from mobile subscribers. The MSC is connected to the backbone wire-line network such as the public

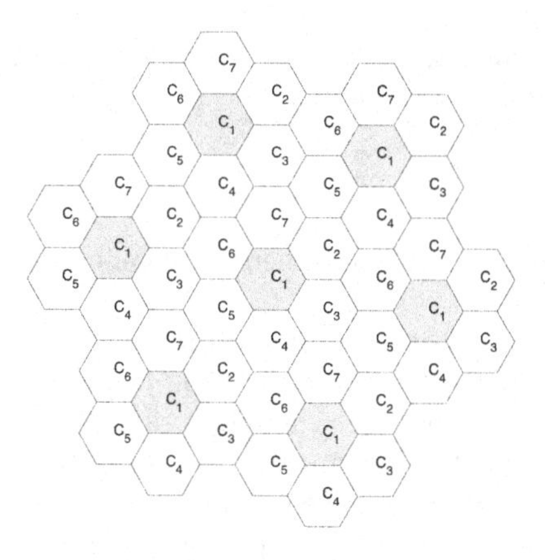

Figure 1.1: Clustered cells in a cellular structure

switched telephone network (PSTN), integrated series digital network (ISDN) or any LAN-WAN-based network. The MSC is also connected to a location database, which stores information about the location of each mobile terminal. The base station is responsible for the communication between the mobile terminal and the rest of the information network. It can communicate with mobiles as long as these are within its operating range. The range itself depends upon the transmission power of the base station. The number of simultaneous calls a mobile wireless system can accommodate is essentially determined by the total spectral allocation for that system and the bandwidth required for transmitting signals in handling a call.

We shall now describe how the use of cells can increase the capacity of a wireless system, allowing more users to communicate simultaneously. The number of simultaneous calls a mobile wireless system can accommodate is essentially determined by the total spectral allocation for that system and the bandwidth required for transmitting signals used in handling a call. This is the same for other radio applications such as broadcast radio, AM or FM. If one considers the first-generation analogue mobile system in the US (the AMPS system), 25 MHz of spectrum is made available for each direction of transmission in the 800-900 MHz radio band. In making radio spectrum allocations for mobile radio communication, the Federal Communications Commission (FCC) in 1981 assigned the 824-849 MHz band for the uplink or reverse channel communication, mobile to base station; the 869-894 MHz was assigned to down-link or forward communication for BS to mobile. Frequency modulation (FM) was the modulation type adopted for these first-generation systems, with the 25 MHz band in each direction broken into signal channels that are 30 KHz wide, each accommodating one call. There are thus 832 analogue signal channels made available in each direction. The 832 analogue channels, or some multiple thereof using digital technology, are obviously insufficient to handle massive numbers of users mainly in urban/suburban areas. Dividing a region into a number of geographically distinct areas called cells and reusing the frequencies in these cells allows the number of communication channels to be increased. The idea of having cellular systems is to reuse channels in different cells, thus increasing the capacity. However, the same frequency assignments cannot be made in adjacent cells because of inter-channel interference. The assignments must be spaced far enough apart geographically to keep interference at tolerable levels. Channel reuse is thus not as efficient as might be expected, but the use of a large number of cells does provide an overall gain in system capacity - that is, the ability to handle simultaneous numbers of calls. In particular, if cells can be reduced in size, more of them can be added in a given geographical area, increasing the overall capacity. The recent trend is to use smaller cells (micro-cells).

To show how the introduction of cells has improved the system's capacity, we consider a one-dimensional case with a first-generation analogue system as an example. Suppose that the overall band of 832 channels is first divided into four groups of 208 channels each. We label these groups as 1, 2, 3, 4 and their locations are shown in Figure 1.2. There are three separate cells with the

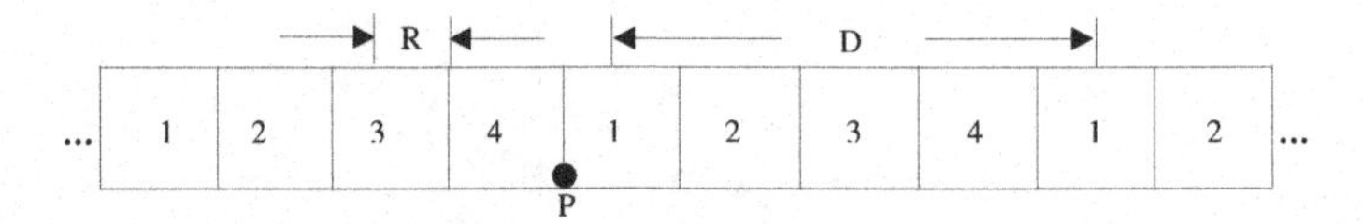

Figure 1.2: A one-dimensional cellular array

same set of frequencies, so we can call this a four-cell reuse. So, given N cells in a system, $208N$ channels are made available compared to the original 832 possibilities when no cellular structure is introduced. With the number N of cells large enough, a significant increase in system capacity has been made possible over the original 832 channels. Had only two cells been required to separate cells using the same band of frequencies (three-cell reuse), a system with N cells would result in $277N$ usable channels, bringing an even larger improvement in capacity. The assignment strategy to be used depends on the tolerable interference, since spacing same-band cells three cells apart results in less interference than spacing them into two cells apart.

Bibliography

Beckmann, D. & Killat, U. (1999). A new strategy for the application of genetic algorithms to the channel-assignment problem. *IEEE Transactions on Vehicular Technology* 48(4), 1261–1269

Castelino, D., Hurley, S. & Stephens, N. (1996). A tabu search algorithm for frequency assignment. *Annals of Operations Research* 63, 301–319

Chen, H., Zeng, Q. & Agrawal, D. (2002). A novel analytical model for optimal channel partitioning in the next generation integrated wireless and mobile networks. Proceedings of the 5th ACM International Workshop on Modelling Analysis and Simulation of Wireless and Mobile Systems. Atlanta, Georgia. pp. 120–127

Chen, H., Zeng, Q.-A. & Agrawal, D. (2003). Evaluation of a new adaptive resource management scheme for multiclass wireless and mobile networks. *in* Proceedings of the Vehicular Technology Conference, 2003. VTC 2003-Fall. 2003 IEEE 58th. Orlando, Florida, USA. pp. 2187–2191

Creput, J., Koukam, A., Lissajoux, T. & Caminada, A. (2005). Automatic mesh generation for mobile network dimensioning using evolutionary approach. *IEEE Transactions on Evolutionary Computation* 9(1), 18–30

Funabiki, N. & Takefuji, Y. (1992). A neural network parallel algorithm for channel assignment problems in cellular radio networks. *IEEE Transactions on Vehicular Technology* 41(4), 430–437

Gamst, A. & Rave, W. (1982). On frequency assignment in mobile automatic telephone systems. Proceedings of the IEEE Global Communication Conference (GLOBECOM. 82). pp. 309–315

Ghosh, S., Sinha, B. & Das, N. (2003). Channel assignment using genetic algorithm based on geometry symmetry. *IEEE Transactions on Vehicular Technology* 52(4), 860–875

Ghosh, S., Sinha, B. & Das, N. (2006). Coalesced CAP: An improved technique for frequency assignment in cellular networks. *IEEE Transactions on Vehicular Technology* 55(2), 640–653

Hale, W. (1980). Frequency assignment: Theory and applications. Proceedings of IEEE. Vol. 68. pp. 1497–1514

Hao, J. & Perrier, L. (1996). Tabu-search for the frequency assignment problem in cellular radio networks. French Workshop on Practical Solving of NP-Complete Problems. Dijon, France

Hao, J., Dorne, R. & Galinier, P. (1998). Tabu search for frequency assignment in mobile radio networks. *J. Heuristics* 4, 47–62

Kendall, G. & Mohamad, M. (2004). Solving the fixed channel assignment problem using an adaptive local search. Proceedings of the 5 th International Conference on Practice & Theory of Automated Timetabling. Pittsburg, USA. pp. 219–231

Kim, J., Park, S., Dowd, P. & Nasrabadi, N. (1996). Channel assignment in cellular radio using genetic algorithms. *Wireless Personal Communications* 3, 273–286

Kim, J., Park, S., Dowd, P. & Nasrabadi, N. (1997). Cellular radio channel assignment using a modified Hopfield network. *IEEE Transactions on Vehicular Technology* 46(4), 957–967

Kim, S., Smith, A. & Lee, J. (2007). A memetic algorithm for channel assignment in wireless FDMA systems. *Computers and Operations Research* 34, 1842–1856.

Kunz, D. (1991). Channel-assignment for cellular radio using neural networks. *IEEE Transactions on Vehicular Technology* 40(1), 188–193

Lai, W. (1996). Channel assignment through evolutionary optimisation. *IEEE Transactions on Vehicular Technology* 45(1), 91–96

Lázaro, O. & Girma, D. (2001). Impact of decentralisation on the performance of a Hopfield neural network-based DCA scheme. *IEEE Communications Letters* 5(11), 444–446

Li, S. & Wang, L. (2001). Channel assignment for mobile communications using stochastic chaotic simulated annealing. *Lecture Notes in Computer Science* 2084, 757–764

Lima, M., Araujo, A. & Cesar, A. (2002). Dynamic channel assignment in mobile communications based on genetic algorithms. Proceedings of the 13th IEEE International Symposium on Personnal, Indoor and Mobile Radio Communication, 15-18 September, 2204–2208

MacDonald, V. (1979). The cellular concept. *The Bell System Technical Journal* 58(1), 15–41

Metzger, B. (1970). Spectrum management technique. 38th national ORSA meeting, 389–397

Nie, J. & Haykin, S. (1999*a*). A dynamic channel assignment policy through Q-learning. *IEEE Transactions on Neural Networks* 10(6), 1443–1455

Nie, J. & Haykin, S. (1999*b*). A Q-learning-based dynamic channel assignment technique for mobile communication systems. *IEEE Transactions on Vehicular Technology* 48(5), 1676–1687

Sandalidis, H., Strvroulakis, P. & Rodriguez-Tellez, J. (1998). An efficient evolutionary algorithm for channel resource management in cellular mobile systems. *IEEE Transactions on Vehicular Technology* 2(4), 125–137

Santos, V., Dinis, M. & Neves, J. (2001). Inclusion of optimisation methods on a new dynamic channel allocation scheme. In F. d Foz, ed., 'Confetele'. Portugal, 581–585

Sarkar, S. & Sivaranjan, K. (1998). Channel assignment algorithms satisfying co-channel and adjacent channel reuse constraints in cellular mobile networks. *IEEE Transactions on Vehicular Technology* 51(5), 954–967

Sivaranjan, K., McEliece, R. & Ketchum, J. (1989). Channel assignment in cellular radio. Vehicular Technology, IEEE Conference-VTC-Spring,vol 2 846–850

Vidyarthi, G., Ngom, A. & Stojmenovic, I. (2005). A hybrid channel assignment approach using an efficient evolutionary strategy in wireless mobile networks. *IEEE Transactions on Vehicular Technology* 54(5), 1887–1895

Wang, W. & Rushforth, C. (1996). An adaptive local search algorithm for the channel assignment problem (CAP). *IEEE Transactions on Vehicular Technology* 45(3), 459–466

Wong, S. (2003). Channel allocation for broadband fixed wireless access networks. Unpublished PhD thesis. University of Cambridge. UK.

Yen, K. & Hanzo, L. (2004). Genetic algorithm assisted multiuser detection in asynchronous CDMA communications. *IEEE Transactions on Vehicular Technology* 53(5), 1413–1422

Yeung, K. L. (2000). Fixed channel assignment optimisation for cellular mobile networks. *IEICE Transactions on Communications* E83-B(8), 1783–1791

Yue, W. (1991). Analytical methods to calculate the performance of a cellular mobile radio communication system with hybrid channel assignment. *IEEE Transactions on Vehicular Technology* VT-40(2), 453–459

Chapter 2

The channel allocation problem

2.1 Channel allocation

A given radio spectrum (or bandwidth) can be divided into a set of disjoint or non-interfering radio channels. All such channels can be used simultaneously while still maintaining an acceptable received radio signal. Different methods exist to divide a given radio spectrum, these being frequency division (FD), time division (TD) and code division (CD). In frequency division, the spectrum is divided into disjoint frequency bands. In time division, the channel separation is achieved by dividing the usage of the channel into disjoint time periods called time slots. In code division, the channel separation is achieved by using different modulation codes. More elaborate techniques can be used to divide a radio spectrum into a set of disjoint channels based on a combination of the above techniques. For example, a combination of TD and FD can be used by dividing each frequency band of an FD scheme into time slots. The major driving factor in determining the number of channels with certain quality that can be used for a given wireless spectrum is the level of received signal quality that can be achieved in each channel.

In order to establish communication with a base station, a mobile terminal must first obtain a channel from the base station. A channel consists of a pair of frequencies: one frequency (the forward link/down link) for transmission from the base station to the mobile terminal and another frequency (the reverse link/up link) for transmission in the reverse direction. An allocated channel is released under two scenarios: the user completes the call or the mobile user moves to another cell before the call is completed. The capacity of a cellular system can be described in terms of the number of available channels, or the number of users the system can support. The total number of channels made available to a system depends on the allocated spectrum and the bandwidth of each channel. Due to limited availability of frequency spectrum and an increasing number of mobile users daily, the channels must be reused as much as possible in order to increase the system capacity. The assignment of channels to cells or mobiles is one of the fundamental resource management issues in a mobile communication system. The channel assignment problem was first introduced in Metzger (1970). The role of a channel assignment scheme is to allocate channels to cells or mobiles in such a way so as to minimise the probability that the incoming calls are blocked, ongoing calls are dropped, and the carrier-to-interference ratio of any call falls below a pre-specified threshold.

2.1.1 Interference

Radio transmission is such that the transmission in one channel causes interference with other channels. Such interference may degrade the signal quality and the quality of service. The potential types of radio interference to a call are:

1. Co-channel interference (c.c.c)- this interference is due to the allocation of the same channel

to certain pair of cells close enough to cause interference (i.e. a pair of cells within the reuse distance);

2. Adjacent channel interference (a.c.c)- this interference is due to the allocation of adjacent channels (e.g f_i and f_{i+1}) to certain pairs of cells simultaneously; and

3. Co-site interferences (c.s.c)- this interference is due to the allocation of channels in the same cell that are not separated by some minimum spectral distance.

These constraints are known as Electromagnetic Compatibility Constraints and can be represented by a minimum channel separation between any pair of channels assigned to a pair of cells or a cell itself. If there are f channels to serve n cells in the system, the minimum channel separation required for an acceptable level of interference is described by a symmetric compatibility matrix C. Each element $c_{i,j}$ $(i, j = 1, ..., n)$ represents the minimum separation required between channels assigned to cells i and j for an acceptable level of interference. (Gamst and Rave, 1982) defined the general form of the CAP in an arbitrary inhomogeneous cellular radio network.

The reuse of channels in cellular systems is inevitable; on the other hand one has to take care of the co-channel interference, thus all channels may not be reused in every cell. However, the concept of a cellular system enables the discrete channels assigned to a specific cell to be reused in different cells separated by a distance sufficient to bring the value of co-channel interference to a tolerable level, thereby reusing each channel many times. The minimum distance required between the centres of two cells using the same channel to maintain the desired signal quality is known as the reuse distance (D_s). The cells with centre-to-centre distance of less than D_s belong to the same cluster within which no channels are allowed to be reused.

Channel allocation is one of many ways to reduce interference in a cellular network. Reduced interference leads to an increase in capacity and throughput of the system. Hence good channel allocation results in a more effective use of the frequency spectrum. Apart from reducing interference, channel allocation algorithms can also be used to adapt to traffic changes in a network, and together with reduced interference, the traffic that can be supported is higher. The task of channel allocation in a cellular system is to allocate c available channels to B base stations, where each base station i has a specific traffic demand that requires d_i channels. Hence, channel allocation is a permutation problem with a search space of

$$\frac{D!}{(D-c)!},$$

where D is the total traffic demand of all the B base stations. In other words $D = \sum d_i$, where d_i represents the number of frequencies assigned to cell i.

To illustrate the channel allocation problem, we consider a simple cellular network with three cells, each one with a base station, and let the set of base stations $B = \{b_1, b_2, b_3\}$ as shown in Figure 2.1 (Wong, 2003). The cellular system requires a channel separation of at least 2 for calls within the same cell and 0 for calls in neighbourhood cells. We therefore have the compatibility matrix as follows:

$$C = \begin{pmatrix} 2 & 0 & 0 \\ 0 & 2 & 0 \\ 0 & 0 & 2 \end{pmatrix} \tag{2.1}$$

Let us assume that the traffic demand $D = [2\,1\,1]$ and the set of channels is given as $\{f_1, f_2, f_3\}$. Define a solution $A = [a_{ij}]$ where

$$a_{ij} = \begin{cases} 1, & \text{if channel } j \text{ is assigned to base station } i, \\ 0, & \text{otherwise.} \end{cases} \tag{2.2}$$

A possible solution is

$$\begin{pmatrix} 1 & 0 & 0 \\ 0 & 1 & 0 \\ 0 & 0 & 1 \end{pmatrix} \tag{2.3}$$

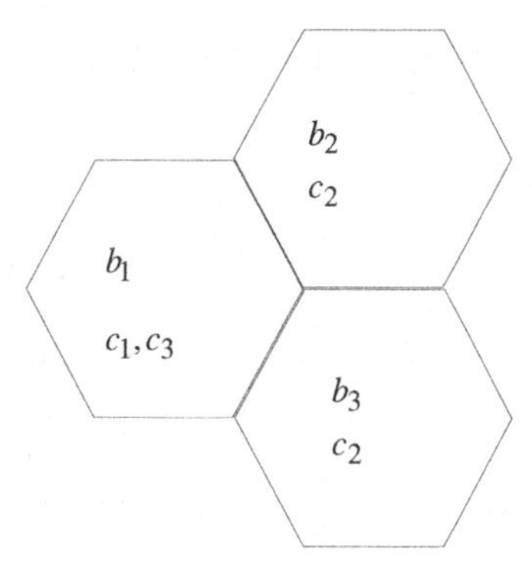

Figure 2.1: Three cellular structure

2.2 Formulation of the CAP

The channel allocation problem in cellular networks has been modelled as an optimisation problem with binary solutions. The problem is characterised by a number of cells and a number of channels, n and c respectively. The solution to the problem is represented by a matrix A. Each element of the matrix is defined according to the equation (2.2).

A general form of matrix A is shown in Figure 2.2. Given the compatibility matrix C and the demand vector D, the aim of the channel allocation problem in the cellular network is to find a conflict-free assignment with the minimum number of frequencies.

2.3 Channel allocation schemes

Channel allocation schemes can be divided into a number of different categories depending on the comparison basis. For example, when channel algorithms are compared based on the manner in which co-channels are separated, they can be divided into three main categories:

1. Fixed channel allocation (FCA) (Cox and Reudink, 1972*a*),
2. Dynamic channel allocation (DCA) (Cox and Reudink, 1972*b*),
3. Hybrid channel allocation (HCA) (Kahwa and Georganas, 1978).

In FCA schemes, the area is partitioned into a number of cells and a number of channels assigned on the desired signal quality. Since the channels in FCA are static, it is not easy to adapt to changes in interference and traffic conditions. As mentioned earlier, in order to overcome those deficiencies of FCA schemes, DCA strategies have been introduced. In DCA, all channels are placed in a pool and they are assigned to new calls as needed, contingent upon a set of conditions (e.g. compatibility matrix) being satisfied. Unlike FCA, the base stations employing DCA do not own any particular channels and a channel is released when a call is completed. At the cost of higher complexity, DCA schemes provide flexibility and traffic adaptability. Although DCA performs better than FCA under high to moderate traffic, its performance is worse than that of FCA under conditions of heavy traffic changes (Katzela & Naghshineh, 1996). To take advantage of both strategies, HCA techniques have been designed. HCA schemes are a mixture of FCA and DCA techniques. In HCA strategies, channels are grouped into two sets; one set of channels is statistically assigned to a base station as in FCA and the second set is placed in a central pool, assigned in a DCA manner.

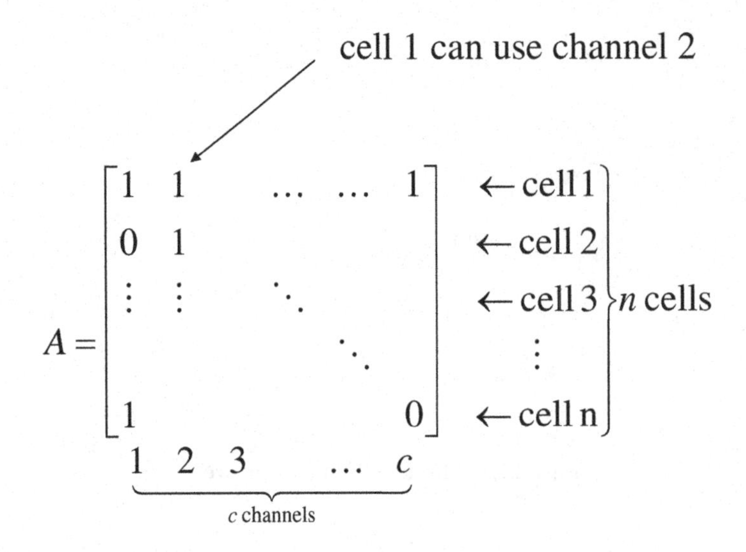

Figure 2.2: Structure of the allocation matrix A

2.3.1 Fixed channel allocation

In the FCA strategy a set of nominal channels is permanently allocated to each cell for its exclusive use. A definite relationship is assumed between each channel and each cell, in accordance with co-channel reuse constraints. The total number of available channels in a system is divided into sets, and the minimum number of channel sets N_c required to serve the entire coverage area is related to the reuse distance as defined by equation (1.1).

In a simple FCA strategy, the same number of nominal channels is allocated to each cell. This uniform channel distribution is efficient if the traffic distribution of the system is also uniform. In this case, the overall blocking probability of the system is the same as the call blocking probability in a cell. Since cellular systems can be non-uniform with temporal and spatial fluctuations, a uniform allocation of channels to cells may result in high blocking in some cells, while others may have a sizeable number of redundant channels. This may result in poor channel utilisation.

In non-uniform channel allocation the number of nominal channels allocated to each cell depends on the expected traffic profile in that cell. Therefore, heavily loaded cells are assigned more channels than lightly loaded ones. Channels are allocated to cells in such a way that the number of blocked calls of the entire system is minimised.

2.3.2 Dynamic channel allocation

Due to short-term temporal and spatial variations of traffic in cellular systems, FCA schemes are not able to attain high channel efficiency. To overcome this, DCA schemes have been studied, and in contrast to FCA, there is no fixed relationship between channels and cells in DCA. All channels are kept in a central pool and are designed dynamically to radio cells as new calls arrive in the system. After a call is completed, its channel is returned to the central pool.

In DCA, a channel is eligible for use in any cell provided that signal interference constraints are satisfied. In general, more than one channel might be available in the central pool to be assigned to select a cell that requires a channel. Thus some strategy must be used to assign channels and the main idea in DCA schemes is to evaluate the cost of using each candidate. The one which gives the optimal solution is chosen. The cost function varies from DCA schemes. Some examples of cost functions (Tekinay and Jabbari, 1991):

- The future blocking probability in the vicinity of the cell;
- The usage frequency of the candidate channel;
- The reuse distance;
- Channel occupancy distribution under current traffic conditions;
- Radio channel measurements of individual mobile users; and
- The average blocking probability of the system.

Different DCA schemes exist, such as centralised DCA schemes and distributed DCA. These are discussed in more detail in Katzela & Naghshineh (1996).

2.3.3 Comparison between FCA and DCA

As mentioned earlier, FCA schemes are more suitable to mainly high uniform traffic load whereas DCA schemes perform better under low traffic density in case of non-uniform traffic load. Therefore, there is a trade-off between quality of service, the implementation complexity of the channel allocation algorithms and spectrum utilisation efficiency. In the FCA schemes channels are preassigned to cells, so there may be instances of blocked calls when there is fluctuation in traffic even though there are available channels in adjacent cells.

The initiation of requests for service from cell to cell is a random process. Therefore, when dynamic assignment is used, different channels are assigned to serve calls at random too. Due to this randomness, it is found that cells which have borrowed the same channel for use are, on average, spaced a greater distance apart than the minimum reuse distance. Consequently, dynamic assignment schemes are not always successful in reusing the channels the maximum possible number of times. On the other hand, in FCA schemes a specific channel can be assigned to cells that are a minimum distance apart such that no interference occurs. The assignment is done in such a way that the maximum reusability of channels is always achieved and this is the main reason that FCA performs better in heavy load traffic. Everitt (1990) has shown that in the case of DCA schemes, the system is not overly sensitive to time and spatial changes in offered traffic, thus giving rise to almost stable performance in each cell. Furthermore, the quality of service within an interference group of cells depends on the average loading within that group, not on its spatial distribution. On the other hand, for the FCA, the service deviation, a measure of the grade of service fluctuations from one cell to another, is very much worsened by time and spatial traffic changes.

Normally, for the same blocking rate, DCA has a lower forced call termination rate than FCA. In FCA a call must be handed off to another channel at every handoff because the same channel is not available in adjacent cells. In DCA, the same channel can be assigned in the new cell if co-channel interference does not occur. In micro-cellular systems, mobiles cross cell boundaries frequently and the traffic of each cell varies drastically. Application of DCA in such a system will perform better due to the fact that DCA adapts to flexibility in the system. It has been shown in Okada and Kubota (1991) that FCA does not give the best result when the cells are small.

In the FCA scheme, the assignment control is made independently in each cell by selecting a vacant channel among those allocated to that cell in advance. On the other hand, in DCA, a knowledge of occupied channels in the relevant cell and also the other cells is needed. Thus DCA requires a lot, and complete knowledge of the state of the entire system and needs high-speed computing and signalling, otherwise there would be a long call set-up delay. In fact, the physical implementation of DCA requires a lot of processing time to determine optimal allocations. But on the other hand, FCA requires a good and labour-intensive frequency planning to start a system, whereas in DCA these are not required. The performance comparison of the two schemes are summarised in Table 2.1 (Katzela & Naghshineh, 1996).

2.3.4 Hybrid channel allocation

In HCA, the total number of channels available for service is divided into fixed and dynamic sets. In the fixed sets, channels are assigned using FCA schemes and the second set of channels is shared by all users in the system. If a call requires service from a cell and all the channels from the fixed sets have been exhausted, then a channel from the dynamic set is allocated to the call using the DCA scheme. A call blocking probability in an HCA scheme is defined as the probability that a call arrives to a cell and finds both the fixed and dynamic channels busy. The ratio of fixed to dynamic channels is a key parameter which determines the performance of the system. The ratio of fixed to dynamic channels is a function of the traffic load and would vary over time according to offered load distribution estimations.

Table 2.1: Comparison between FCA and DCA

FCA	DCA
Performs better under heavy traffic	Performs better under light/moderate traffic
Low flexibility in channel assignment	Flexible allocation of channels
Maximum channel reusability	Not always maximum reusability
Sensitive to time and spatial changes	Insensitive to time and time spatial changes
Not stable grade of service per cell in an interference cell group	Stable grade of service per cell in an interference cell group
High forced call termination probability	Low to moderate forced call termination probability
Suitable for large cell environment	Suitable in micro-cellular environment
Low flexibility	High flexibility
Radio equipment covers all channels assigned to cells	Radio equipment covers the temporary channels assigned to the cell
Independent channel-control fully centralised to fully distributed	Control dependent on the scheme
Low computational effort	High computational effort
Low call set-up delay	Moderate to high call set-up delay
Low implementation complexity	Moderate to high implementation complexity
Complex, labour-intensive frequency planning	No frequency planning
Low signalling load	Moderate to high signalling load
Centralised control	Centralised, decentralised, distributed control depending on the scheme

Bibliography

Cox, D. & Reudink, D. (1972*a*). A comparison of some channel assignment strategies in large mobile communication systems. *IEEE Transactions on Communications* 20, 190–195

Cox, D. & Reudink, D. (1972*b*). Dynamic channel assignment in two-dimension large-scale mobile radio systems. *Bell System Technical Journal* 51(7), 1611–1628

Everitt, D. (1990). Traffic capacity of cellular mobile communications systems. *Computer Networks and ISDN Systems* 20, 447–454

Gamst, A. & Rave, W. (1982). On frequency assignment in mobile automatic telephone systems. In 'Proceedings of the IEEE Global Communication Conference (GLOBECOM'82)'. pp. 309–315

Kahwa, T. & Georganas, N. (1978). A hybrid channel assignment scheme in large-scale, cellular structured mobile communications systems. *IEEE Transactions on Communications* COM-26(4), 432–438

Katzela, I. & Naghshineh, M. (1996). Channel assignment schemes for cellular mobile telecommunications systems. *IEEE Personnal Communication* pp. 10–31

Metzger, B. (1970). Spectrum management technique. *National ORSA meeting* **38th**, 389–397

Okada, K. & Kubota, F. (1991). On dynamic channel assignment in cellular mobile radio systems. *in* 'Proceedings of the IEEE International Symposium on Circuits and Systems'. Vol. 2. pp. 938–941

Tekinay, S. & Jabbari, B. (1991). Handover and channel assignment in mobile cellular networks. *IEEE Communications Magazine*

Wong, S. (2003). Channel allocation for broadband fixed wireless access networks. Unpublished PhD thesis. University of Cambridge. UK

Chapter 3

Mathematical formulation of the CAP

3.1 Introduction

This chapter describes the mathematical formulation associated with each of the schemes that we have described for the CAP. Frequencies or channels are represented by the positive integers 1, 2, 3,..., f, where f is the maximum allocation of the spectrum bandwidth. The basic model of the channel allocation problem for the different schemes is presented below.

3.1.1 Problem formulation – FCA

The basic model of the channel assignment problem can be represented as follows (Sivaranjan et al., 1989; Wang and Rushforth, 1996):

1. n: the number of cells in the network;
2. d_i: the number of frequencies required in cell i $(1 \leq i \leq n)$ in order to satisfy channel demand;
3. C: compatibility matrix, $C = (c_{ij})$ denotes the frequency separation required between cell i and cell j; and
4. f_{ik_i}: a radio channel is assigned to k^{th} call in cell i.

The objective of the CAP is

$$\min_{i,k_i} f_{ik_i}, \tag{3.1}$$

subject to

$$|f_{ik_i} - f_{jk_j}| \geq c_{ij},\ \forall\ i,\ j,\ k_i \neq k_j.$$

The channel allocation problem in the cellular network is finding a conflict-free assignment with the minimum number of total frequencies, where C, the compatibility matrix and D, the demand vector are given. In other words, one tries to find the minimum of

$$\max_{ik} f_{ik}.$$

3.1.2 Problem formulation – DCA

The DCA scheme which aims to reduce the effect of unbalanced loading due to evenly distributed traffic sources, assigns channels to cells on a call-to-call basis.
Notation (Sivaranjan et al., 1990):

1. n: the number of cells in the system;
2. C: compatibility matrix, $C = (c_{ij})$ denotes the frequency separation required between cell i and cell j;
3. n_i, $1 \leq i \leq n$: the number of calls in progress in cell i;
4. p_i, $1 \leq i \leq n$: the probability that a new call arrives in cell i;
5. ρ: the total traffic in the system.
6. $\rho_i = p_i\rho$, $1 \leq i \leq n$: the traffic in cell i;
7. N_f: the number of (contiguous) frequency channels available. These channels are numbered 1 through N_f; and
8. f_{ik_i}, $1 \leq i \leq n$, $1 \leq k_i \leq m_i$: the frequency assigned to the kth call in the ith cell.

Constraints: $|f_{ik_i} - f_{jl_j}| \geq c_{ij}$ for all i, j, k_i, l_j except for $i = j$, $k_i = l_j$.
Assumptions:

- Call arrivals in cell i are independent of all other arrivals and obey a Poisson distribution with parameter ρ_i;
- Call holding times are exponentially distributed with mean call duration.
- There are no calls handed off between cells; and
- Blocked calls are cleared (Erlang B).

3.1.3 Problem formulation – HCA

The total number of channels available for service is divided into fixed and dynamic sets. The fixed set contains a number of nominal channels that are assigned to cells as in the FCA schemes and, in all cases, are to be preferred for use in their respective cells. The dynamic set is shared by all users in the system to increase flexibility.

A request for a channel from the dynamic set is initiated only when the cell has exhausted using of all its channels from the fixed set.

3.1.4 An example of a four-cellular network

The following example of a four-cellular network that is $N = 4$ (Funabiki and Takefuji, 1992; Sivaranjan et al., 1989; Smith and Palaniswami, 1997) has compatibility matrix C given by

$$C = \begin{pmatrix} 5 & 4 & 0 & 0 \\ 4 & 5 & 0 & 1 \\ 0 & 0 & 5 & 2 \\ 0 & 1 & 2 & 5 \end{pmatrix}. \tag{3.2}$$

Let $D = (d_i)_{1 \leq i \leq 4}$ denote the demand vector [1 1 1 3]. The assignment to be generated is denoted by an $n \times c$ binary matrix A, whose element a_{ij} is 1 if channel j is assigned to cell i, and 0 otherwise as shown below. This implies that the total number of 1s in row i of matrix A must be d_i, and a possible structure of matrix A is given as

$$A = \begin{pmatrix} 1 & 0 & 0 & 0 & 0 & 0 & 0 & 0 & 0 & 0 & 0 & 0 & 0 \\ 0 & 0 & 0 & 0 & 1 & 0 & 0 & 0 & 0 & 0 & 0 & 0 & 0 \\ 1 & 0 & 0 & 0 & 0 & 0 & 0 & 0 & 0 & 0 & 0 & 0 & 0 \\ 1 & 0 & 0 & 0 & 0 & 0 & 0 & 1 & 0 & 0 & 0 & 0 & 1 \end{pmatrix}. \quad (3.3)$$

We now illustrate by means of an example the notations used in the problem formulation.

1. The number of frequencies/radio channels required in each cell is given in Table 3.1.

Table 3.1: Channel demand

Cell	Cell 1	Cell 2	Cell 3	Cell 4
Number of Channels	1	1	1	3

2. Minimum frequency separation required between cell i and cell j.

Table 3.2: Constraints based on compatibility matrix C

	Cell 1	Cell 2	Cell 3	Cell 4
Min freq separation dist	Co-site :5 Cell 2/4	Co-site :5 Cell 1/4 Cell 4/1	Co-site :5 Cell 4/2	Co-site :5 Cell 3/2 Cell 2/1

Notation: The co-site constraint shows the minimum separation between two frequencies assigned to the same cell. i/j indicates that for the cell to which that column refers, there must be a separation of j between that cell and cell i.

3. The call list that needs to be assigned to each cell j.

Table 3.3: Call list

	Cell 1	Cell 2	Cell 3	Cell 4
Call list	$Call_{11}$ $(i = 1,\ k = 1)$	$Call_{21}$ $(i = 2, k = 1)$	$Call_{31}$ $(i = 3, k = 1)$	$Call_{41}$, $Call_{42}$, $Call_{43}$ $(i = 4, 1 \leq k \leq 3)$

4. A frequency is assigned to serve a call in each cell.

Table 3.4: Frequency at every cell

	Cell 1	Cell 2	Cell 3	Cell 4
Call list	1 call	1 call	1 call	3 calls
frequency f_{ik}	f_{11}	f_{21}	f_{31}	f_{41}, f_{42}, f_{43}

5. Frequency separation constraint.

6. The total frequency that is to be assigned

$$TotalAssignCh = (d_1 + d_2 + d_3 + d_4) = 6.$$

Table 3.5: Frequency channel separation constraint C, where $k \geq 1$, $m \leq 3$

	Cell 1	Cell 2	Cell 3	Cell 4
Cell 1	–	$\|f_{21} - f_{11}\| \geq 4$	–	–
Cell 2	$\|f_{11} - f_{21}\| \geq 4$	–	–	$\|f_{4k} - f_{21}\| \geq 1$
Cell 3	–	–	–	$\|f_{4k} - f_{31}\| \geq 2$
Cell 4	–	$\|f_{21} - f_{4k}\| \geq 1$	$\|f_{31} - f_{4k}\| \geq 2$	$\|f_{4k}\text{-}f_{4m}\| \geq 5$

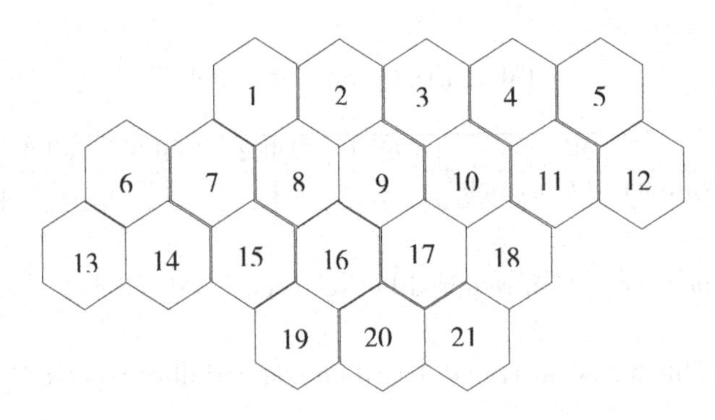

Figure 3.1: Structure of cellular systems

3.2 Lower bounds for the CAP

Several authors (Gamst, 1986; Smith and Hurley, 1997) have studied lower bounds for the CAP with reference to the Philadelphia problem. A structure of the cellular system of the Philadelphia problem is shown in Figure 3.1. Lower bounds for the frequency assignment problem can be used to assess the performance of the various methods. One of the first papers found was by Gamst (1986) in which the best bounds for the CAP were presented. Smith and Hurley (1997) and Tcha et al. (1997) have also presented bounds for the frequency assignment problem. In Tcha et al. (1997), the work done was an extension of Gamst (1986). The procedure devised for frequency insertion made the best of unexploited frequency spaces between the assigned frequencies. In Sung and Wong (1995) and Sung and Wong (1997), tighter lower bounds than those presented in Gamst (1986) were derived. Later, Janssen and Kilakos (1999) and Smith et al. (2000) made use of some mathematical programming techniques and new lower bounds were presented. Ghosh et al. (2003) have found some lower bounds for cellular networks with homogeneous demands. Table 3.7 shows the tightest bounds found for some most common benchmark problems for the demand vectors D_1 and D_2 as given in Table 3.6. Mainly the simulations will be based on these benchmark problems defined in Table 3.6 and our results will be compared to the lower bounds presented in Table 3.7. As will be seen later in this book, many approaches to solve the CAP have been used and the aim was to achieve the theoretical lower bounds under the constraints defined by equation (3.1). Though there are several benchmark problems, the Philadelphia with 21 cells network problems have been the most commonly used, hence our motivation for testing our methodology with them.

3.3 Summary

In this chapter we have described the cellular networks and introduced the different channel allocation schemes, namely the FCA, DCA and the HCA and their mathematical formulation. A

Table 3.6: Demand vectors D_1 and D_2

	d_1	d_2	d_3	d_4	d_5	d_6	d_7	d_8	d_9	d_{10}	d_{11}
D_1	8	25	8	8	8	15	18	52	77	28	13
D_2	5	5	5	8	12	25	30	25	30	40	40
	d_{12}	d_{13}	d_{14}	d_{15}	d_{16}	d_{17}	d_{18}	d_{19}	d_{20}	d_{21}	
D_1	15	31	15	36	57	28	8	10	13	8	
D_2	45	20	30	25	15	15	30	20	20	25	

Table 3.7: Interference constraints

Problem	1	2	3	4	5	6	7	8
a.c.c	1	2	1	2	1	2	1	2
c.s.c	5	5	7	7	5	5	7	7
D_1/D_2	D_1	D_1	D_1	D_1	D_2	D_2	D_2	D_2
Lower Bound	381	427	533	533	221	253	309	309

comparison between the different channel allocation schemes is made and mathematical formulations of the FCA and DCA schemes are given. A simple four-channel example is included so that one can understand with the compatibility matrix and the demand vector how channels are allocated. Furthermore, a structure of the classical Philadelphia was presented together with theoretical lower bounds and the respective demand vectors.

Bibliography

Funabiki, N. & Takefuji, Y. (1992). A neural network parallel algorithm for channel assignment problems in cellular radio networks. *IEEE Transactions on Vehicular Technology* 41(4), 430–437

Gamst, A. (1986). Some lower bounds for a class of frequency assignment problems. *IEEE Transactions on Vehicular Technology* VT- 35(1), 8–14

Ghosh, S., Sinha, B. & Das, N. (2003). Channel assignment using genetic algorithm based on geometry symmetry. *IEEE Transactions on Vehicular Technology* 52(4), 860–875

Hale, W. (1980). Frequency assignment: Theory and applications. In 'Proceedings of IEEE'. Vol. 68. pp. 1497–1514

Janssen, J. & Kilakos, K. (1999). An optimal solution to the 'philadelphia' channel assignment problem. *IEEE Transactions on Vehicular Technology* 48(3), 1012–1014

Janssen, J., Kilakos, K. & Marcotte, O. (1999). Fixed preference channel assignment for cellular telephone systems. *IEEE Transactions on Vehicular Technology* 48(2), 533–541

Montemanni, R., Moon, J. & Smith, D. (2003). An improved tabu search algorithm for the fixed spectrum frequency assignment problem. *IEEE Transactions on Vehicular Technology* 52(3), 891–901

Revuelta, L. (2007). A new adaptive genetic algorithm for fixed channel assignment. *Information Sciences* 177, 2655–2678

Sivaranjan, K., McEliece, R. & Ketchum, J. (1989). Channel assignment in cellular radio. Vol.2. VTC 89. pp. 846–850

Sivaranjan, K., McEliece, R. & Ketchum, J. (1990). Dynamic channel assignment in cellular radio. *Procedings of the IEEE* 40th *Vehicular Technology Conference* pp. 631–637

Smith, D. & Hurley, S. (1997). Bounds for the frequency assignment problem. *Discrete Mathematics* 167/168, 571–582

Smith, D., Hurley, S. & Allen, S. (2000). A new lower bound for the channel assignment problem. *IEEE Transactions on Vehicular Technology* 49(4), 1265–1272

Smith, K. & Palaniswami, M. (1997). Static and dynamic channel assignment using neural networks. *IEEE Journal on Selected Areas in Communications* 15(2), 238–249

Sung, C. & Wong, W. (1995). A graph theoretic approach to the channel assignment problem in cellular systems. In 'Proceedings of the IEEE Vehicular Technology Conference'. Vol(2). VTC 95, 604–608

Sung, C. & Wong, W. (1997). Sequential packing algorithm for channel assignment under co-channel and adjacent channel interference constraint. *IEEE Transactions on Vehicular Technology* 46(3), 676–686

Tcha, D., Chung, Y. & Choi, T. (1997). A new lower bound for the frequency assignment problem. *IEEE/ACM Transactions on Networking* 5(1), 34–39

Wang, L. & Tian, F. (2000). Noisy chaotic neural networks for solving combinatorial optimisation problems. In Proceedings of the International Joint Conference on Neural Networks.4, 37–40

Wang, W. & Rushforth, C. (1996). An adaptive local search algorithm for the channel assignment problem (CAP). *IEEE Transactions on Vehicular Technology* 45(3), 459–466.

Audhya, G., Sinha, K., Ghosh, S. & Sinha, B. (2011). A survey on the channel assignment problem in wireless networks. *Wireless Communications and Mobile Computing and other Emerging Wireless Network Technologies* 11(5), 583–609.

Part II
Methods used to solve channel allocation problem

This part contains a review of the different algorithms used in solving the CAP since the early 1970s. In fact, a survey of a large number of published papers in the area of the channel allocation problem was done to compile all the information given in the following chapters. The algorithms can be broadly categorised in terms of local search, simulated annealing, graph-theory-based methods, neural networks, genetic algorithms and tabu search. A summary of the work done before 1980 can be found in Hale (1980), but since then no compilation of published papers has been done. Thus we find that such a survey is important so that we can gather the strengths and weaknesses of each method used. However, a recent paper by Audhya et al. (2011) has been found, but their approach is different from our work.

Chapter 4

Graph theory

4.1 Graph colouring approach

Graph colouring is an assignment of colours to elements of a graph subject to certain constraints. It is a way of colouring the vertices of a graph such that no two adjacent vertices share the same colour. Graph colouring has many applications such as scheduling (Marx, 2004), sudoku, mobile radio frequency assignment, pattern matching and register allocation(Chaitin, 1982), as well as theoretical challenges.

The problem discussed in Chapter 3 can be modelled as a graph colouring problem in the following way. The vertices V of graph G consist of a set of calls to be assigned. In our case they are $call_{11}$, $call_{21}$, $call_{31}$ and $call_{41}$, $call_{42}$, $call_{43}$. An edge, $E \rightarrow e(call_{ij}, call_{kl})$ for a graph G can be defined as a constraint between $call_{ij}$ and $call_{kl}$, shown in Figure 4.1. The aim is to schedule the calls such that each 'clashed' call will be assigned a different channel. In the concept of graph colouring, the channel is represented by a colour and the objective is to minimise the number of colours used.

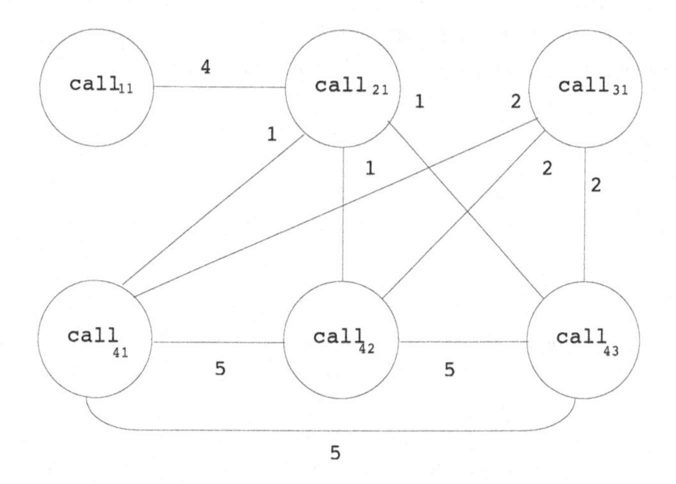

Figure 4.1: Example of a four-cell graphical representation

In general, there are two objectives in solving the channel allocation problem:

- Given a traffic demand, base station number and compatibility matrix, we find the number of frequency channels without any interference constraints; in other words, we minimise the total bandwidth (span) of radio channels such that traffic demand and interference constraints are satisfied.

- Given a number of frequencies, a number of base stations, traffic demand and compatibility matrix, we minimise channel interferences such that demand constraints are satisfied.

The basic idea of the graph theoretic approach is to list the calls in some order and use either a frequency exhaustive assignment strategy (FEA) or a requirement exhaustive assignment strategy (REA). The FEA strategy starts with the first cell in the ordered list and each call is assigned the least possible frequency, consistent with previous assignment, that is, without violating the separation constraints. A pseudo-code of FEA is given in Algorithm 4.1.

Algorithm 4.1 Frequency exhaustive assignment strategy

1 - Loop 1: over all call i in the call list
2 - Loop 2: for each channel j within the lower bound
3 - If i is not assigned and j can be assigned to i
4 - Assign j to i
5 - Break Loop 2
6 - End If
7 - End Loop 2
8 - End Loop 1

In REA, an attempt is made to assign the first channel to all cells which have unsatisfied channel requirements, starting from the first cell.

Algorithm 4.2 Requirement exhaustive assignment strategy

1 - Loop 1: for each channel j within the lower bound
2 - Loop 2: for each call i in the call list
3 - If i is not assigned and j can be assigned to i
4 - Assign j to i
5 - Break Loop 2
6 - End If
7 - End Loop 2
8 - End Loop 1

Then the same procedure is followed to assign the remaining calls. The procedure is repeated until all the channel requirements are exhausted. We present a pseudo-code for the REA in Algorithm 4.2.

Metzger (1970) recognised that the classical node-colouring problem in graph theory is analogous to frequency assignment problems where only co-channel constraints are involved. A graph is a set of nodes partially or completely interconnected by lines. In colouring a graph, adjacent nodes may not receive the same colour. The objective is to find the minimum number of colours required to colour all the nodes. In the frequency assignment case, the nodes represent requirements and the lines connect pairs of requirements that cannot be assigned to the same channel, and the colours represent the channels.

Metzger (1970) employed decomposition procedures which are described as follows:

- A node of lowest degree from the graph along the lines connected to it is removed;
- Removal of a node of lowest degree in the reduced graph; and
- Procedure is repeated until the last node is removed from the graph.

The concept of the node colouring order technique by Metzger has shown to be an important technique in solving the CAP using the graph colouring approach. Another colouring procedure in which the distribution of the number of colours used is made as even as possible is known as the uniform assignment strategy.

In Zoeller and Beall (1977), the authors have used three different assignment strategies: frequency exhaustive, requirement exhaustive and uniform assignment with three different order techniques, which are node colouring, node degree and random order. The degree of a cell i is defined as

$$deg_i = \sum_{j=1}^{n} d_i c_{ij}. \tag{4.1}$$

The node degree order arranges the cells in decreasing order of their degrees and the node colouring order is as follows: of the n cells, the cell with the least degree is placed at the last nth place list. This cell is eliminated from the system and the degrees of the remaining cells are recomputed. Now, the cell with the least degree is placed at the $(n-1)$th position in the list, and eliminated from the system. This process is continued until the ordering is complete.

These different procedures were evaluated for the following cases: co-channel, adjacent and co-site cases. The results obtained by Zoeller and Beall (1977) show that about 35% more spectrum may have been committed to existing requirements than is really needed. This represents a significant under-utilisation of the spectrum. Upwards of 20-25% of the spectrum might be recovered in reassigning present users with the use of node-colouring order-based assignment procedure.

Gamst and Rave (1982) summarised four existing sequential algorithms. The first algorithm has four different versions by combining two different assignment strategies, namely the frequency exhaustive assignment and the requirement exhaustive assignment, and two different order strategies given by the node degree order and the node colouring order (Zoeller and Beall, 1977). The second algorithm repeatedly assigns frequencies according to the assignment difficulty of requirements (Box, 1978). The third algorithm uses the heuristic geometric principle of maximum overlap of denial areas. It states that a frequency should be assigned to a cell whose denial areas have the maximum overlap with the existing denial area of that frequency. The fourth algorithm is based on graph theory, where the clique number plays a key role. Sivaranjan et al. (1989) proposed an $O(n^2)$ time sequential heuristic algorithm based on the first algorithm introduced by Gamst and Rave (1982). They applied their algorithm to several problems, where the values of total frequencies in solutions are shown without any actual assignment results. Later, Sengoku et al. (1991) formulated a channel offset system design using a graph theoretical concept in which the degree of co-channel interference between cells was represented.

Two algorithms based on graph colouring are reported in Sung and Wong (1997) and later by Janssen et al. (1999). Later, in Yeung (2000), six channel assignment algorithms were presented and evaluated. The methods were the combinations of three channel assignment strategies and two cell ordering methods. The following findings were stated in the paper:

1. The node colour method of ordering cells is a more efficient ordering method than the node-degree ordering method.

2. The frequency exhaustive strategy is more suitable for systems with highly non-uniform distributed traffic, and the requirement exhaustive strategy is more suitable for systems with less non-uniform distributed traffic.

3. The combined frequency and the requirement exhaustive strategy with node colour re-ordering is the most efficient algorithm. The frequency span obtained in Yeung (2000) using the

mentioned algorithms was much lower than that reported in earlier published papers. In many cases, results equal to the theoretical lower bounds were obtained. In Malesinska and Panconesi (1997), a channel stability number has been considered for the hybrid cellular networks using graph theory.

4.2 Summary

For this method, the cellular network is modelled as a graph and the CAP has been formulated as a graph colouring problem. In some of the studies mentioned in this section, the graph used to model the cellular network ignores the geometry of the network whereas others take into account the geometry and they have solved the CAP optimally. Some authors have considered only the co-channel constraints whereas some considered both the co-channel and the adjacent channel constraints.

Bibliography

Box, F. (1978). A heuristic technique for assigning frequencies to mobile radio nets. *IEEE Transactions on Vehicular Technology* VT-27, 57–64

Chaitin, G. J. (1982). Register allocation and spilling via graph colouring. 'SIGPLAN Symposium on Compiler Construction'. pp. 98–105

Gamst, A. & Rave, W. (1982). On frequency assignment in mobile automatic telephone systems. 'Proceedings of the IEEE Global Communication Conference (GLOBECOM'82)'. pp. 309–315

Janssen, J., Kilakos, K. & Marcotte, O. (1999). Fixed preference channel assignment for cellular telephone systems. *IEEE Transactions on Vehicular Technology* 48(2), 533–541

Malesinska, E. & Panconesi, A. (1997). On the hardness of allocating frequencies for hybrid networks. *Theoretical Computer Science* 209, 347–363

Marx, D. (2004). Graph coulouring problems and their applications in scheduling. *Periodica Polytechnica, Electrical Engineering* 48, 11–16

Metzger, B. (1970). Spectrum management technique. *National ORSA meeting* 38th, 389–397

Sengoku, M., Tamura, H., Shinoda, S., Abe, T. & Kajitani, Y. (1991). Graph theoretical considerations of channel offset systems in a cellular mobile system. *IEEE Transactions on Vehicular Technology* 40(2), 405–411

Sivaranjan, K., McEliece, R. & Ketchum, J. (1989). Channel assignment in cellular radio. Vol. VTC 89. pp. 846–850

Sung, C. & Wong, W. (1997). Sequential packing algorithm for channel assignment under co-channel and adjacent channel interference constraint. *IEEE Transactions on Vehicular Technology* 46(3), 676–686

Yeung, K. L. (2000). Fixed channel assignment optimisation for cellular mobile networks. *IEICE Transactions on Communications* E83-B(8), 1783–1791

Zoeller, J. A. & Beall, C. L. (1977). A breakthrough in spectrum conserving frequency assignment technology. *IEEE Transactions on Electromagnetic Compatibility* EMC 19, 313–319

Chapter 5

Neural networks

5.1 Neural network approach

A neural network model is composed of a large number of massively connected simple processing elements. The processing element is called a neuron because it performs the function of a simplified biological neuron model and it has inputs and an output. The input of a processing element is connected with outputs of several processing elements, including the processing element itself. The interconnections between neurons are given by the motion equation:

$$\frac{dU_i}{dt} = \frac{\partial E(V_1,\ V_2,\ ...,\ V_n)}{\partial V_i}, \tag{5.1}$$

where U_i and V_i are respectively the input and the output of neuron i, and n is the number of required neurons for the problem. The computational energy function $E(V_1,\ V_2,\ ...,\ V_n)$ describes all the constraints and the goal function of the problem. A quadratic form of non-negative function is usually adopted as the energy function. The motion equation performs the gradient descent method to minimise the energy function. The output state vector $(V_1,\ V_2,\ ...,\ V_n)$ of neurons represents a solution of the problem when the energy function is minimised.

The neural network model for solving the combinatorial optimisation problem was first introduced by Hopfield & Tank (1986). An energy function, E, represents the distance between the current state of the neural network system and the solution state of the neural network system, and it is determined by considering all constraints in the problem. The goal of the neural network model for solving combinatorial optimisation problems is to minimise the energy function E.

The application of neural network in the channel allocation problem was first proposed by Sengoku et al. (1991) and Kunz (1991). The latter did not consider the adjacent channel constraint, and limited the co-site constraint to $c_{ii} = 2$. He used the continuous Hopfield network, but the method had some disadvantages due to the use of the slow sigmoid neuron model ($f(x) = 1/2(1 + \tanh(\lambda x))$) and the harmful decay term in the motion equation. Furthermore, a careful tuning of the coefficients in the motion equation is required for each different problem and a careful gain control of the sigmoid function has to be provided in order to obtain valid solutions. Kunz did not discuss the convergence frequency of the neural network system which is always controversial in neural network research.

Funabiki and Takefuji (1992) proposed the neural network for the general case. The proposed parallel algorithm was based on an artificial neural network composed of nm processing elements for an n-cell and an m-frequency problem. The McCulloch-Pitts neuron model (McCulloch and Pitts, 1943) was used in the paper. It has been shown empirically that the McCulloch-Pitts neuron model provides faster convergence to the solution than the sigmoid neuron model used in Kunz (1991) and also improves the convergence frequency by suppressing the undesirable oscillatory behaviour of the McCulloch-Pitts neuron model. In the Funabiki and Takefuji (1992) model, four heuristics were used to improve the convergence rate of channel assignment. This proposed parallel algorithm was tested on eight benchmark problems. They showed that their method allowed the simulations to start with the lower bounds. In order to accelerate the convergence time in some problems, the frequency assignment was fixed in one or more cells or certain cells with the largest number of required frequencies. In fact, fixing a single cell frequency assignment can drastically reduce the search space and consequently the convergence time is shortened. The simulation results were then compared with simulations performed in Kunz (1991). The authors concluded that their proposed algorithm found solutions in nearly constant time, with favourable results in some cases. In both approaches discussed above, the neural network requires an $n \times m$ neurons, where n is the number of cells and m is the number of channels available to the cellular systems. Therefore, for a cellular system which features many cells and channels, a large number of neurons is required and as such they will be inefficient and computationally expensive.

Sengoku et al. (1991) proposed a neuron structure and tested it on a small system consisting of only eight cells. Later, Funabiki and Nishikawa (1995) proposed an improved neural network to solve the CAP and they ran the simulations for much larger systems. They introduced an energy function which considered the constraint that n cells must be assigned one of m channels and that co-channel interference must be avoided. Thus the energy function was given by

$$E = \frac{A}{2}\sum_{i=1}^{n}\left(\sum_{k=1}^{m} V_{ik} - 1\right)^2 + B\sum_{i=1}^{n}\sum_{j=1}^{m} f\left(\sum_{k=1,\, k\neq i}^{n} w_{ik}V_{kj} - I_{max}\right)V_{ij}, \tag{5.2}$$

where the function $f(x) = x$ if $x > 0$, and 0 otherwise, and A and B are constant coefficients. To improve the convergence of the neural network, two heuristics, namely the hill-climbing term (Takefuji and Lee, 1991; Funabiki and Takefuji, 1994) and an omega function, were used.

An algorithm proposed by Funabiki and Nishikawa (1995) is given below as an example of how the neural networks perform. However, different authors have proposed different algorithms.

Algorithm 5.1: Neural network algorithm

---

1 - Set values for parameters t, A, B
2 - Set initial values of $U_{ij}(t)$ and $V_{ij}(t)$
3 - Compute the motion equation given in (5.1)
4 - Update $U_{ij}(t+1)$
5 - Update $V_{ij}(t+1)$
6 - Increment the number of iteration steps t by one
7 - Check termination condition else Go To step 3

Chan et al. (1994) used a feed-forward neural network, which had a learning process prior to actual channel assignment. For the learning process, they used training data that were dependently obtained by other assignment methods. The performance of their algorithm was totally dependent on the used training data and the authors considered only the co-channel constraint.

In Kim et al. (1995, 1997), the authors have used a modified Hopfield hhh neural network. The latter used a forced assignment and the changing cell list order technique to inhibit falling into the local minima, which is one of the disadvantages of neural networks. In the algorithm, the interconnection weights between the neurons are designed in such a way that each neuron received inhibitory support if constraint conditions were violated and received excitatory support if the constraint conditions were satisfied. The interconnection weight was set initially to take into account the required channel number in each cell and the three channel assignment constraints. An initialisation method was used as the initial states for neurons in a Hopfield network. Two initialisation techniques, which used the specific characteristics of frequency assignment problems such as the co-site interference, adjacent channel interference and the co-channel interference constraints, were considered so as to accelerate the convergence rate. The major difference between the algorithm presented in Kim et al. (1995) and in Kim et al. (1997), was that in the previous algorithm, some frequencies were fixed so as the convergence time was accelerated, whereas in the latter, no frequency was fixed before the frequency assignment procedure. The results reported in the latter were better than reported in the previous paper by the same author.

Smith and Palaniswami (1997) proposed two different neural networks to address the FCA and DCA problem. The first method was an improved neural network which resolved the issues of infeasibility and poor solution quality and the second was a self-organising neural network. The improved Hopfield network proposed by the authors was able to eliminate the problem of infeasibility and poor solution quality, by expressing all constraints of the problem in a single term and modifying the internal dynamics of the Hopfield network. The advantage of the energy function is that only one penalty term needs to be selected and if this term was large enough then the validity of the solution is ensured. Furthermore, the modification of the internal dynamics allowed temporary increases of the energy function so as to allow an escape from the local minima. The major difference to the Hopfield method was that a feed-forward neural network was used to minimise the interference. The networks were then tested on test problems found in literature and a comparison was made. The proposed networks were also applied to the DCA. In this paper, the CAP has been reformulated as a generalised quadratic assignment problem, treating the noninterference constraints as soft constraints and the demand satisfaction as hard constraints.

In Wilmes and Erickson (1996), two methods of dynamic channel allocation using neural network are investigated. Neural networks were used to learn trends that occur in calling traffic networks and future traffic prediction was then used to manage resources accordingly. The first method used a back-propagation model prediction to identify spatial traffic patterns as a function of the time during the day. The future traffic predictions were fed into the channel allocator in fixed discrete time increments so that the channel could uncover emerging traffic trends over a short period of time and allocate resources accordingly. The second method was built on the first method by feeding the back-propagation predictions into a real-time learner, which learnt a history of traffic activity between two adjacent cells, and then made the allocation decision.

Fernandes and da Silva (2001) have solved the CAP by minimising the interference. For this purpose two neural networks have been used, mainly the Hopfield and the self-organising neural networks. The objective function used in the optimisation algorithms had two terms, namely an interference term and a penalty term. The interference term forced the algorithms to minimise the interference, while the penalty term forced the demanded channels to be assigned. The same problem formulation used in Smith and Palaniswami (1997) has been used in this paper. The interference term was defined with the help of a tensor P obtained from the interference matrix with the following recursive law:

$$\begin{aligned} P_{j,i,1} &= C_{ij},\ \forall\, j,\ i \neq j, \\ P_{j,i,m+1} &= \max(P_{j,i,m-1}, 0),\ m = 1,\ ...,\ j-1, \\ P_{j,j,1} &= 0,\ \forall\, j = 1,\ ...,\ n. \end{aligned} \tag{5.3}$$

The optimisation problem was expressed in terms of the objective function with constraints, as follows:

$$\begin{aligned} \min\ F(X) &= \sum_{j=1}^{n}\sum_{k=1}^{f} A_{j,k} \sum_{i=1}^{n}\sum_{l=1}^{f} P_{j,i(|k-l|+1)} A_{i,l}, \\ \sum_{k=1}^{f} d_j &= D_j,\ j = 1,\ \ldots,\ n, \\ A_{j,k} &\in \{0,1\},\ \forall j,\ k, \end{aligned} \tag{5.4}$$

and the energy function was given as

$$E = -\frac{1}{2}x^T W x - \theta^T x, \tag{5.5}$$

which is in the form of a quadratic optimisation problem. Here W denotes the weight matrix of a Hopfield network of n units, x is the state and θ is the n-dimensional vector of units thresholds. The Hopfield neural network (Smith and Palaniswami (1997)) was used to solve the problem. Defining Δt as the convergence step, the following three methods, which were used to decrease execution time, were described:

- Hopfield network with constant $\triangle t$;
- Hopfield network with variable $\triangle t$, dependent on the energy function. Faster progression of the energy function increased the convergence step, while a slower progression (close to minimum) decreased $\triangle t$; and
- Hopfield network with variable $\triangle t$ dependent on the number of iterations of the algorithm. As the number of iterations increases, $\triangle t$ decreases.

In order to provide a local minimum escape method to the Hopfield network, a new $\alpha(t)$ was coupled. This term allowed the control of the descent and climb of the energy function, providing a means to escape from local minima. A self-organising neural network was also used. The major difference to the Hopfield method is that a feed-forward neural network is used to minimise interference. To evaluate the performance of the algorithms, two kinds of simulations were made, namely FCA and the DCA. The FCA simulations were performed on a series of test problems described in literature, whereas the DCA simulations were performed for a small problem having four cells and 11 channels. The results show that the Hopfield network is a faster algorithm.

Lázaro and Girma (2000) considered the application of a Hopfield neural network to dynamic channel allocation. They have modelled and examined the effect on performance of traffic mobility and the consequent inter-cell call handoff, which, under increasing load, can force call terminations with an adverse impact on quality of service; thus forced call termination should be kept to a minimum. In order to meet those objectives, a new energy function was formulated to cater for new call arrival and handoff channel allocation due to traffic mobility. The energy function was a modification of the one proposed by Re et al. (1996), with the only difference being the last term of the equation (5.6).

$$\begin{aligned} E &= \frac{A}{2}\sum_{j=1}^{M}\sum_{i=1,\, i\neq i^*}^{n}(V_{i^*,j}A_{i,j}c_{ii^*}) + \frac{B}{2}\left[\sum_{j=1}^{n}(V_{i^*,j} - Traf(i^*))\right]^2 \\ &- \frac{C}{2}\sum_{j=1}^{M}\sum_{i=1,\, i\neq i^*}^{n}\left(V_{i^*,j}A_{ij}\frac{1-c_{ii^*}}{Dist(i,i^*)}\right) - \frac{D}{2}\sum_{j=1}^{n}(V_{i^*,j}A_{i^*,j}) \\ &+ \frac{F}{2}\sum_{j=1}^{C}\sum_{i=1,\, i\neq i^*}^{n}\Big(V_{i^*,j}A_{ij}[1 - Res(i,i^*)]\Big) - \frac{G}{2}\sum_{j=1}^{C}[Free_j(1 - V_{i^*,j} - H).] \end{aligned} \tag{5.6}$$

To define the symbols which have been used in the above equation:
M: Total number of channels in the system.
H: Total number of handoff guard channels.
i^*: Cell involved in a call arrival or dropping.
$Traf(i^*)$: Number of requested channels at cell i^*.
$Dist(i,i^*)$: Distance between cell i and i^* normalised to adjacent cells separation.
$Res(i,i^*)$: Element of a reuse matrix whose value is 1 if cells i and i^* belong to same reuse pattern, and is 0 otherwise.

$V_{i^*,j}$: Assignment of the cell of interest.
$Free_j$: The elements of a free channel vector $Free$ whose value is 1 if channel j is not being used by any cell in the system and 0 otherwise.

The essence of the new energy function formulation was in incorporating a channel reservation control for the ongoing calls, so that handoff channel allocations had a degree of priority over incoming calls. The three new parameters G, $Free_j$ and H were the value-added terms in the energy function; in fact, they tended to restrict channels being assigned to new calls if the demand could be fulfilled with those already allocated to the system. Hence this constraint ensured that some channels were always reserved for handoff calls. Three different neural networks were considered:

- HNN DCA;
- HNN DCA+: The HNN DCA augmented with a prioritised call handling for handoff channel assignment; and
- HNN DCA++: The one with the new formulation as described.

Simulations were performed under both uniform and non-uniform traffic conditions. It was shown that the blocking probability of HNN DCA could be dropped by a priority scheme and that results for HNN DCA++ were much better. An important point is that there was no call blocking with HNN DCA++ in non-uniform traffic conditions. In Lázaro and Girma (2001), important aspects of a distributed cellular optimisation technique based on a simulation model of a Hopfield-neural-network-based dynamic allocation scheme is evaluated. Distributed schemes can be applied to reduce the optimisation area and hence the problem size, thereby speeding up computation time and distributing channel assignment signalling, while at the same time providing a comparable level of performance to that of centralised schemes. The effectiveness of the decentralised control relied on knowledge sharing of the various optimisation areas, which was facilitated by the provision of signalling connectivity for channel allocation messaging. Based on the proposed decentralisation technique and evaluation, changing the distributed optimisation area size clearly provided an added adaptation mechanism for optimising signalling load and computation time of a micro-cellular DCA system without significant performance degradation.

Zhenya et al. (2002) presented a multistage self-organising channel assignment algorithm with a transiently chaotic neural network as an optimisation technique. The advantage of this algorithm lay in the chaotic neuro-dynamics which were temporarily generated for searching and self-organising in order to escape local minima. Thus the neural network gradually approached, through transient chaos, a dynamical structure similar to conventional models such as Hopfield neural networks. Recently Wang et al. (2005) proposed a discrete competitive Hopfield neural network (DCHNN) to solve the channel allocation problem. The authors have shown that the DCHNN could always guarantee the feasibility of the solutions for the CAP. Furthermore, the DCHNN allowed temporary energy increases to escape from local minima by the introduction of stochastic dynamics. Simulations were performed on Kunz's data and showed that the proposed method had superior ability for the CAP within reasonable number of iterations.

The Hopfield neural network has also been applied to the DCA in Lázaro and Girma (1999, 2000) and to borrowing channel allocation in Sandalidis et al. (1998).

5.2 Summary

Neural-network-based approaches use an energy function which contains the objective function as well as an individual term for each of the constraints of the problem. The terms of the energy function compete among themselves for minimisation. The Hopfield neural network has been the most commonly used network in solving the CAP. The main objective is to minimise the total interference in the whole cellular network structure. The main disadvantage of the neural network approach is that it tends to converge to local optima, and thus optimal solutions cannot be guaranteed.

Bibliography

Chan, P., Palaniswami, M. & Everitt, D. (1994). Neural network based dynamic channel assignment for cellular mobile communications systems. *IEEE Transactions on Vehicular Technology* 43(2), 279–288

Fernandes, T. R. C. C & da Silva, H. J. A (2001). Solving the channel assignment problem using neural networks and genetic algorithms. 'Proceedings of the Conference on Telecommunication–ConfTele'. Vol. 01. Figueira da Foz, Portugal. pp. 297–301

Funabiki, N. & Nishikawa, N. (1995). An improved neural network for channel assignment in cellular mobile communication systems. *IEICE Transactions on Communications* E78-B(8), 1187–1195

Funabiki, N. & Takefuji, Y. (1992). A neural network parallel algorithm for channel assignment problems in cellular radio networks. *IEEE Transactions on Vehicular Technology* 41(4), 430–437

Funabiki, N. & Takefuji, Y. (1994). A parallel algorithm for time-slot assignment problems in TDM hierarchical switching systems. *IEEE Transactions on Communications* 42(10), 2890–2898

Hopfield, J. & Tank, D. (1986). Computing with neural circuits: A model. *Science* 233, 625–633

Kim, J., Park, S., Dowd, P. & Nasrabadi, N. (1995). Comparison of two optimisation techniques for channel assignment in cellular radio network. *in* 'Proceedings of the IEEE International Conference Communications'. pp. 1850–1854

Kim, J., Park, S., Dowd, P. & Nasrabadi, N. (1997). Cellular radio channel assignment using a modified Hopfield network. *IEEE Transactions on Vehicular Technology* 46(4), 957–967

Kunz, D. (1991). Channel-assignment for cellular radio using neural networks. *IEEE Transactions on Vehicular Technology* 40(1), 188–193

Lázaro, O. & Girma, D. (1999). Dynamic channel allocation based on a Hopfield neural network and requirements for autonomous operation in a distributed environment. *in* 'Proceedings of the IEEE 50th Vehicular Technology Conference'. Vol. VTC - 4. pp. 2334–2338

Lázaro, O. & Girma, D. (2000). A Hopfield neural-network based dynamic channel allocation with handoff channel reservation control. *IEEE Transactions on Vehicular Technology* 49(5), 1578–1587

Lázaro, O. & Girma, D. (2001). Impact of decentralisation on the performance of a hopfield neural network-based DCA scheme. *IEEE Communications Letters* 5(11), 444–446

McCulloch & Pitts, W. (1943). A logical calculus of ideas immanent in nervous activity. *Bulletin of Mathematical Biophysics* 5(4), 115–133

Re, E., Fantacci, R. & Ronga, L. (1996). A dynamic channel allocation technique based on hopfield neural networks. *IEEE Transactions on Vehicular Technology* 45(1), 26–32

Sandalidis, H., Strvroulakis, P. & Rodriguez-Tellez, J. (1998). An efficient evolutionary algorithm for channel resource management in cellular mobile systems. *IEEE Transactions on Vehicular Technology* 2(4), 125–137

Sengoku, M., Tamura, H., Shinoda, S., Abe, T. & Kajitani, Y. (1991). Graph theoretical considerations of channel offset systems in a cellular mobile system. *IEEE Transactions on Vehicular Technology* 40(2), 405–411

Smith, K. & Palaniswami, M. (1997). Static and dynamic channel assignment using neural networks. *IEEE Journal on Selected Areas in Communications* 15(2), 238–249

Takefuji, Y. & Lee, K. C. (1991). Artificial neural networks for four coloring map problems and k-colorability problems. *IEEE Transactions on Circuits Systems* 38(3), 326–333

Wang, J., Tang, Z., Xu, X. & Li, Y. (2005). A discrete competitive Hopfield neural network for cellular channel assignment problems. *Neurocomputing* 67, 436–442

Wilmes, E. & Erickson, K. (1996). Two methods of neural network controlled dynamic channel allocation for mobile radio systems. *in* 'Proceedings of the IEEE Vehicular Technology Conference'. Vol. 2. pp. 746–750

Zhenya, H., Zhang, Y., Wei, C. & Wang, J. (2002). A multistage self-organizing algorithm combined transiently chaotic neural network for cellular channel assignment. *IEEE Transactions on Vehicular Technology* 51(6), 1386–1396

Chapter 6

Heuristics

6.1 Heuristic techniques

A heuristic is a technique that gives near-optimal solutions at reasonable computational cost without being able to guarantee either feasibility or optimality or to state how close to optimality a particular feasible solution is. Heuristics techniques are non-algorithmic methods that are applied to algorithmically complex or time-consuming problems that do not have a pre-determined methods for generating efficient solutions. There is no analytic methodology to explain the way the heuristic converges on a solution; this is achieved with the partial control of some external factors. Hence heuristics are often said to be guided random search methods. Heuristics have been suggested to solve a wide range of problems in various fields including artificial intelligence and continuous and discrete combinatorial optimisation Reeves (1995).

CAP has been solved by heuristic methods and Box (1978) proposed an automated heuristic assignment technique in which channel requirements that proved themselves to be difficult, were solved. The heuristic technique was capable of solving complex channel assignment problems that involve co-channel, adjacent channel, spurious and intermodulation interference; non-repetitive zone structures; fixed pre-existing frequency assignments; and frequency resource lists that contain gaps and are varied from zone to zone. The heuristic technique was used to solve both FCA and DCA problems. Borrowing channel assignment (BCA) was introduced in literature as a compromise between the classic fixed and dynamic channel allocation schemes. Lower bounds for the frequency assignment problem could be found from maximal cliques and subgraphs related to the cliques. Smith et al. (1998) have shown that the optimal assignment of the CAP could be found by a process of assigning these subgraphs first, fixing the assignment and then extending the assignment to the full problem. The simulation results were performed using the software FASoft (Hurley et al., 1997) The authors concluded that their methods presented in the paper gave better results than a direct application of a heuristic. In Sandalidis et al. (1999) three heuristic BCA techniques were presented, one based on a Hopfield neural network was discussed; an efficient evolutionary algorithm namely combinatorial evolution strategy (CES) and a third heuristic which combined the advantages of the above two methods.

Funabiki (2000) presented a three-stage algorithm heuristic combined neural network for the CAP. The three-stage algorithm consisted of :

1. The regular interval assignment stage;
2. The greedy assignment stage; and
3. The neural network assignment stage.

In the first stage, the calls in a cell determining the lower bound on the total number of channels were assigned channels at regular intervals. In the second stage, the calls in a cell with the largest degree and its adjacent cells were assigned channels by a greedy heuristic method. In the third stage, the calls in the remaining cells were assigned channels by a binary neural network from Funabiki and Takefuji (1992). When the simulation results were compared to benchmark problems, mainly Sivaranjan's benchmark problems, the lower bounds were attained in all the cases in reasonable computational time. The heuristic algorithm of Chakraborty (2001) generated a population of random valid solutions during which the optimisation criterion of minimising the bandwidth was not taken into consideration. Finally the best solutions were the optimal or very near optimal solutions in a very small number of trials.

In Battiti et al. (2001) a new version of the heuristic technique known as randomised saturation degree (RSD), based on node ordering and randomisation, was presented. The solutions obtained by RSD were further improved by means of a local search technique. Experimental results had shown that the RSD method was effective in terms of both solution quality and computational times. Kendall and Mohamad (2004,*b*) presented a hyper-heuristic approach. The method is problem-independent and unlike other meta-heuristic approaches such as variable neighbourhood search, one could move from one neighbourhood to another neighbourhood structure. The difference in Kendall and Mohamad (2004*b*) was that a greedy constructive heuristic was used to generate an initial solution and then applied the method in Kendall and Mohamad (2004) to improve the quality of the solution. A pseudo-code of the hyper-heuristic algorithm used in Kendall and Mohamad (2004) is given below.

Let S be a solution space, f be an objective function and N be a neighbourhood.

Algorithm 6.1 Hyper-heuristic algorithm

---

1 - Initialisation
- 1.1 - Choose an initial solution $s_0 \in S$
- 1.2 - Compute an initial objective function $f(s_0)$
- 1.3 - Set Initial $LEVEL = f(s_0)$
- 1.4 - Set DownRate value

2 - Operation and termination
- 2.1 - Call LLH to generate new neighbour solution $s_n \in N(s_0)$
- 2.2 - Compute neighbour objective function $f(s_n)$
- 2.3 - if $f(s_n) \leq LEVEL$, accept $f(s_0)$ then update $s_0 = s_n$
- 2.4 - Reduce $LEVEL = LEVEL - DownRate$
- 2.5 - If stopping condition = false, Go To step 2.1

Two greedy local search algorithms, namely CAP1 and CAP2, were described. The two algorithms differed in the way in which neighbours were generated. CAP1 algorithm generated neighbours randomly, whereas CAP2 algorithm did so accordingly to a deterministic rule that is in a systematic order. Wang and Rushforth also presented a third algorithm, CAP3, which was a blend of the probabilistic and deterministic neighbour generation. In the computational stage, it was shown that CAP3 outperformed both CAP1 and CAP2. Kendall and Mohamad (2004*c*) have presented a greedy local search combined with a Monte Carlo algorithm. The local framework proposed by Wang and Rushforth (1996) was used to define the solution space and objective function. The greedy local search algorithm consisted of two different stages namely, a probabilistic stage and a new neighbour generation stage.

Algorithm 6.2 Calculation of DownRate

1 - LB = Best result
2 - *Iter* = Number of iterations
3 - DownRate = $(f(s_0) - LB)/Iter$

An acceptance criterion, such as the Exponential Monte Carlo with counter (EMCq) (Glover and Laguna, 1997) was chosen in order to escape from local minima. The simulation results were compared with other acceptance criteria such as random descent. The authors proposed ways of refining their results:

1. Use different strategies to generate initial solutions.
2. Use a different neighbour generating strategy such as reinserting the call based on random selection.
3. Implement additional acceptance criteria.

Algorithm 6.3: Low-level Heuristics (LLH)

1 - Sort the channel from the lowest to highest, delete the call with the highest channel assignment, randomly insert at any point and reassign the channel.
2 - Same as 1, but randomly select the call to delete.
3 - Same as 1, but find the best point at which to reassign.
4 - Same as 1, but randomly change the call order starting from the insertion point.

Dynamic channel allocation (DCA) has also been solved using heuristic approaches. Some authors focussed only on DCA problems, such as in Cheng and Chuang (1996) where a simple aggressive least interference algorithm (LIA) was used. The method performed well under different propagation and traffic conditions. On the other hand, Liu et al. (2001) presented the integer programming model for the DCA under space and time varying traffic demand. The number of channels required to satisfy the traffic demand was minimised using a threshold-based decision criterion. A neighbourhood-based search procedure was used for the most recent channel state information to perform a feasible assignment when there is change in demand. The performance of a greedy sequential channel assignment was also examined relative to the spatial distribution of the cells with time-varying demand.

6.1.1 Summary

The majority of the heuristics treat the CAP as a combinatorial optimisation problem and try to minimise a cost function that is based on the assumptions of the designer. Newer hyper-heuristics have been developed. Their main advantage is that they are problem-independent and are thus a preferred tool for use in optimisation problems. Initial solutions are first generated using a greedy constructive heuristic after which low level heuristics are applied to improve the solutions.

6.2 Simulated annealing

Simulated annealing is used to solve a discrete optimisation problem – that is, one has to express the problem as a cost function optimisation problem by defining the configuration space S, the cost function C and the neighbourhood structure N. It is a general method for obtaining approximate solutions to combinatorial optimisation problems. It was originally presented by Kirkpatrick et al. (1983) and Cerny (1985). The algorithm is a generalisation of the iterative improvement of the local search scheme, but can also be considered as a simulation of the physical annealing processes found in nature. Simulated annealing is regarded as an inhomogeneous Markov process, consisting of a sequence of homogeneous chains at each temperature level t. Its transition matrices are then given by $P(t) = G(t)A(t)$. Let s denote an arbitrary solution and s' be the neighbouring solution of s. $g_{ss'}(t)$ denotes the probability to propose configuration s' while being in configuration s and $a_{ss'}(t)$ denotes the corresponding acceptance probability as stated in Duque-Anton et al. (1993).

The pseudo-code for the simulated annealing is given in algorithm 6.4 When the optimisation process is undertaken in a stochastic manner, it is also called stochastic simulated annealing (SSA).

A choice for the cost function as given by Duque-Anton et al. (1993):

$$C(S) = \frac{1}{2}A \cdot \sum_{(i,j),(i',j'),(i,j)\neq(i',j'),|i-i'|\leq c_{jj'}} s_{ij}s_{i'j'} + \frac{1}{2}B\sum_{j}\left(\sum_{i} s_{ij} - traf_j\right)^2 \tag{6.1}$$

Choices for the neighbourhood of a configuration s are produced by performing the following transitions:

- A single flip, that is, just switching on or off one channel i in one cell j; and
- A flip-flop, that is, replacing at cell j one used channel i_1 with one unused i_2.

Algorithm 6.4: Simulated annealing algorithm

---

1 - Initialise(s_{start}, t_0)
2 - $k := 0$.
3 - $s := s_{start}$
4 - Until *Equilibrium reached* do:
  4.1 - Generate s' from $N(s)$.
  4.2 - Metropolis test:
  4.3 - if $\min\{1, exp(-C(s') - C(s))/t_k\}$ >random[0,1)
    then continue with $s := s'$,
  4.4 - else continue with the old s.
5 - If stop criterion valid, stop.
6 - $k := k + 1$.
7 - Calculate t_k.
8 - Go to 4.

Different models have been used by Duque-Anton et al. (1993) and Mathar and Mattfeldt (1993). The latter also investigated several algorithms based on simulated annealing approach and showed that all variants gave good-quality solutions when compared to optimal solutions. When simulated annealing is used to solve an NP-hard problem it is linked to the question of how to control the quality of obtained solutions, which in general is a difficult task to solve in an effective manner. In recent years, considerable interest has been shown in chaotic simulated annealing (CSA) in different fields (Chen and Aihara, 1995; Wang and Tian, 2000; Tian et al., 2000). CSA can search efficiently because of its reduced search spaces. In Li and Wang (2001) a combination of SSA and CSA were combined with stochastic chaotic simulated annealing (SCSA) and applied to the CAP.

6.2.1 Summary

The differences between the various approaches presented by the different authors lie mainly in the representation of the state space, the definition of the neighbourhood of a state, and the choice of the probability function. The SA approach guarantees a global optimal solution asymptotically, but the rate of convergence is rather slow. To solve the CAP the number of constraint violations is regarded as the cost function. Neighbourhood states are determined through selecting a random frequency at a randomly selected transmitter. The number of constraint violations is then calculated for the new state as well as for the current state.

6.3 Tabu search

Tabu search, which was first suggested by Glover (1977), is based on the belief that intelligent searching should embrace more systematic forms of guidance such as memorising and learning. It is a form of neighbourhood search with a set of critical and complementary components. It is a local or neighbourhood search procedure to move iteratively from a solution s to a solution s' in the neighbourhood of s until some stopping criterion has been satisfied.

A typical tabu search algorithm begins with an initial configuration and then proceeds iteratively to visit a series of locally best configurations following the neighbourhood function. At each iteration, a best neighbour is sought to replace the current configuration. From literature it has been found that very few researchers have used the tabu search in the channel allocation problem. The first papers found were by Bouju et al. (1995), Castelino et al. (1996) and Hao and Perrier (1996). Later, Hao et al. (1998) and Capone and Trubian (1999) developed some more results for the CAP using the tabu search. Recently, an improved tabu search algorithm was presented in Montemanni et al. (2003). The algorithm was shown to have better solution quality rather than solution speed.

Algorithm 6.5: Tabu search algorithm

---

1 - Tabu search (Pr)- Pr is an optimisation problem.
2 - $S :=$ randomly generated solution of Pr.
3 - $Best := S$
4 - While (termination criterion not met)
  4.1 - $S :=$ best solution in the neighbourhood of $S*$.
  4.2 - If ($Cost(S) < Cost(Best)$)
  4.3 - $Best := S$
  4.4 - EndIf
  4.5 - Update tabu list.
5 - EndWhile.
6 - Return $Best$.

6.3.1 Summary

Tabu search is a relatively simple technique which can provide considerably improved results by defining proper rules tailored to specific problems. The disadvantage is that the performance is highly dependent on the fine-tuning of several parameters. Most of the algorithms presented in this section are broadly based on this approach. The differences between these approaches lie in the representation of the move, neighbourhood of a move and the way of defining a tabu move. Some authors have shown that they outperform other approaches such as simulated annealing and genetic algorithm (Capone and Trubian, 1999). Montemanni et al. (2003) presented a robust search algorithm with a dynamic tabu list. The algorithm shows that solution quality is a more important criterion than solution speed.

Bibliography

Battiti, R., Bertossi, A. & Cavallaro, D. (2001). A randomized saturation degree heuristic for channel assignment in cellular networks. *IEEE Transactions on Vehicular Technology* 50(2), 364–374

Bouju, A., Boyce, J., Dimitropoulos, C., von Scheidt, G. & Taylor, J. (1995). Tabu search for the radio links frequency assignment problem. In 'Proceedings for the Conference of Applied Decision Technologies: Modern Heuristic Methods. Brunel University, Uxbridge, UK. pp. 233–250

Box, F. (1978). A heuristic technique for assigning frequencies to mobile radio nets. *IEEE Transactions on Vehicular Technology* VTC - 27(2), 57–64

Capone, A. & Trubian, M. (1999). Channel assignment problem in cellular systems: A new model and a tabu search algorithm. *IEEE Transactions on Vehicular Technology* 48(4), 1252–1260

Castelino, D., Hurley, S. & Stephens, N. (1996). A tabu - search algorithm for frequency assignment. *Annals of Operations Research* 63, 301–319

Cerny, V. (1985). Thermodynamical approach to the travelling salesman problem: an efficient simulation algorithm. *J.Opt. Theory Appl.* 45, 41–51

Chakraborty, G. (2001). An efficient heuristic algorithm for channel assignment problem in cellular radio network. *IEEE Transactions on Vehicular Technology* 50(6), 1528–1539

Chen, L. & Aihara, K. (1995). Chaotic simulated annealing by a neural network model with transient chaos. *Neural Networks* 8(6), 915–930

Cheng, M. & Chuang, J. (1996). Performance evaluation of distributed measurement-based dynamic channel assignment in local wireless communications. *IEEE Journal on Selected Areas in Communications* 14(4), 698–710

Duque-Anton, M., Kunz, D. & Ruber, B. (1993). Channel-assignment for cellular radio using simulated annealing. *IEEE Transactions on Vehicular Technology* 42(1), 14–21

Funabiki, N. (2000). A three stage heuristic combined neural network algorithm for channel assignment in cellular mobile systems. *IEEE Transactions on Vehicular Technology* 49(2), 397–403

Funabiki, N. & Takefuji, Y. (1992). A neural network parallel algorithm for channel assignment problems in cellular radio networks. *IEEE Transactions on Vehicular Technology* 41(4), 430–437

Glover, F. (1977). Heuristics for integer programming using surrogate constraints. *Decision Sciences* 8(2), 156–166

Glover, F. & Laguna, M. (1997). *Tabu search.* Kluwer Academic Publisher

Hao, J. & Perrier, L. (1996). Tabu-search for the frequency assignment problem in cellular radio networks. In French Workshop on Practical Solving of NP-Complete Problems. Dijon, France

Hao, J., Dorne, R. & Galinier, P. (1998). Tabu search for frequency assignment in mobile radio networks. *J. Heuristics* 4, 47–62

Hurley, S., Smith, D. & Thiel, S. (1997). FASoft: A software system for discrete channel frequency. *Radio Science* 32, 1921–1939

Kendall, G. & Mohamad, M. (2004*a*). Channel assignment in cellular communications using a great deluge hyper-heuristic. *in* 'Proceedings of the 12 th IEEE International Conference on Networks'. Vol. 2. pp. 769–773

Kendall, G. & Mohamad, M. (2004*b*). Channel assignment optimisation using a hyper-heuristic. In Proceedings of the IEEE Conference on Cybernetics and Intelligent Systems. Singapore. pp. 790–795

Kendall, G. & Mohamad, M. (2004*c*). Solving the fixed channel assignment problem using an adaptive local search. *in* 'Proceedings of the 5 th International Conference on Practice & Theory of Automated Timetabling'. Pittsburg USA. pp. 219–231

Kirkpatrick, S., Gelatt, C. & Vecchi, M. (1983). Optimization by simulated annealing. *Science* 220(4598), 671–680

Li, S. & Wang, L. (2001). Channel assignment for mobile communications using stochastic chaotic simulated annealing. *Lecture Notes in Computer Science* 2084, 757–764

Liu, S., Daniels, K. & Chandra, K. (2001). Channel assignment for time-varying demand. In Proceedings of the Global Telecommunications Conference, GLOBECOM' 01.IEEE'. Vol. 6. pp. 3563–3567

Mathar, R. & Mattfeldt, J. (1993). Channel assignment in cellular radio networks. *IEEE Transactions on Vehicular Technology* 42(4), 647–656

Montemanni, R., Moon, J. & Smith, D. (2003). An improved tabu search algorithm for the fixed spectrum frequency assignment problem. *IEEE Transactions on Vehicular Technology* 52(3), 891–901

Reeves, C. (1995). *Modern Heuristic Techniques for Combinatorial Problems (Advanced Topics in Computer Science).* McGraw Hill

Sandalidis, H., Strvroulakis, P. & Rodriguez-Tellez, J. (1999). Borrowing channel assignment based on heuristic techniques for cellular systems. *IEEE Transactions on Neural Networks* 10(1), 176–181

Smith, D., Hurley, S. & Thiel, S. (1998). Improving heuristics for the frequency assignment problem. *European Journal of Operational Research* 107(1), 76–86

Tian, F., L.Wang & Fu, X. (2000). Solving channel assignment problems for cellular radio networks using transiently chaotic neural networks. In Proceedings of the International Conference on Automation, Robotics and Computer Vision

Wang, L. & Tian, F. (2000). Noisy chaotic neural networks for solving combinatorial optimisation problems. In Proceedings of the International Joint Conference on Neural Networks

Wang, W. & Rushforth, C. (1996). An adaptive local search algorithm for the channel assignment problem (CAP). *IEEE Transactions on Vehicular Technology* 45(3), 459–466

Chapter 7

Evolutionary methods

7.1 Genetic algorithm

A genetic algorithm (GA) is a search and optimisation method which works by mimicking the evolutionary principles and chromosomal processing in natural genetics. A GA starts its search with a random set of solutions, usually coded in binary string structures. Every solution is assigned a fitness which is directly related to the objective function of the search and optimisation problem. Three operators – reproduction, crossover and mutation – are applied, similar to natural genetic operators, to produce a new population. These three operators are applied in each generation until a termination criterion is satisfied. GAs have been successfully used in solving various problems because of their simplicity, global perspective and inherent parallel processing. Bremermann (1958) and Fraser (1957) can be considered as the pioneers of GAs. The algorithm was later considerably developed by Holland (1975) and Goldberg (1989). Since then, GAs have been widely used as a new computational tool for solving optimisation problems. Cuppini (1994); Kim et al. (1995, 1996) and Lai (1996) were among the first papers found in literature to have applied GA to solve the channel allocation problem; one of the most recent paper was written by Revuelta (2007).

The channel assignment problem was formulated as an energy minimisation problem and in this algorithm the cell frequency was not fixed beforehand. Some years earlier, Ngo and Li (1998) and Beckmann and Killat (1999) developed new genetic algorithms. In Beckmann and Killat (1999) a new strategy, known as the Combined Genetic Algorithm (CGA) was developed, whose method was a blend of the well-known frequency exhaustive strategy and genetic algorithm. In CGA new call lists were generated by the genetic algorithm, then frequency exhaustive strategy was used to evaluate the quality of the generated call lists. The authors started the procedure by estimating the lower bound z on bandwidth (Gamst, 1986; Tcha et al., 1997; Sung & Wong, 1997). If the algorithm did not find a solution with z, the latter was increased by one and the CGA algorithm was repeated until a valid solution was reached or until a maximum number of iterations was attained. Thus, in this approach, the computation time will be highly dependent on the proximity of the estimation of the lower bound on bandwidth to its optimal value. Santos et al. (2000, 2001) presented a new dynamic channel allocation, namely simplified maximum packing, and evaluation was done using genetic algorithms. Lima et al. (2002) introduced two new DCA strategies using genetic algorithms:

1. GAL, where the channels previously assigned are kept blocked during the call holding time.

2. GAS, where the calls can be switched to different channels during the connection time.

These two strategies were then compared with existing methods.

Zomaya (2002) devised a GA-based channel borrower which was mainly based on selective channel borrowing but different to (Lim et al., 1999; Yener and Rose, 1997). When a host entered a new cell, a handoff request was made. If there were channels available, the host was assigned immediately to a free channel. If no channel was available, a channel borrowing operation was attempted. A decision had to be taken on which neighbouring cell from which a channel could be borrowed. A GA was used to help to ease channel borrowing decisions that would minimise denial of service and maximise the long-term performance of the network.

Genetic algorithms have been applied to the channel allocation problem from another perspective Ghosh et al. (VTC Fall 2002, 2003). The authors presented new lower bounds on the required bandwidth for a cellular network with homogeneous demand followed by some strategies for assigning channels to cells of the entire cellular network with a homogeneous demand using GA. Then the strategies for homogeneous demand were extended to the cases with non-homogeneous demand vectors. In Ghosh et al. (2003), an algorithm that would solve the CAP in its general form was introduced using the elitist model of genetic algorithm (EGA). For the case of homogeneous demands, EGA was applied to a small subset of nodes and then extended for the entire cellular network ensuring faster convergence. The algorithm was also applicable to the nonhomogeneous demands. A three-stage heuristic combined genetic algorithm strategy was used to solve the CAP in Fu et al. (2003). The three stages were:

- The Regular Interval Assignment stage;
- The Greedy Assignment stage; and
- The Genetic Algorithm Assignment stage.

The algorithm presented was a composition of Funabiki (2000) and Beckmann and Killat (1999). Yen and Hanzo (2000) have presented a genetic algorithm scheme in conjunction with a local search so as to improve the initial guesses using a population of randomly generated possible solutions, while maintaining a high diversity of solutions. It was shown that the GA was capable of converging to the best solution in a short time.

Algorithm 7.1: Genetic Algorithm

---

1 - Set initial parameters (time (t), number of cells (M))
2 - Initial Population
  2.1 - begin
  2.2 - for i = 0 to M-1 do
  2.3 - begin
  2.4 - Generate a random order of the nodes
    in the CAP graph and consider it as a string S_i
  2.5 - end
  2.6 - Set $q_t \leftarrow \{S_0, S_1, ..., S_{M-1}\}$ and the initial population
  2.7 - end
3 - Compute Fitness function $Fit(S_i)$ for each string S_i of q_t
4 - Reproduction: Apply the selection operation as described below
  on the strings of q_t to generate a mating pool q_{mat} of size M
5 - Crossover: Perform a crossover operation on the strings
  of q_{mat} to obtain a population q_{temp1} of size M
6 - Mutation: Perform a mutation operation on the strings of
  q_{temp1} to obtain a population q_{temp2} of size M
7 - Calculate $Fit(S_i)$ for each string of q_{temp2}
8 - Elitism operation
9 - $t \leftarrow t + 1$; Go To Step 3 it $t < T$, else terminate

---

Wong and Wassell (2002*a*,*b*) applied a novel DCA method using a genetic algorithm for a TDD broadband fixed wireless access network. These authors have categorised the existing channel allocation methods in terms of a Channel Allocation Matrix (Wong, 2003). Genetic algorithms have also been applied in the Assisted Multiuser Detection in synchronous and asynchronous CDMA communications (Juntti, 1997; Ergun and Hacioglu, 2000; Yen and Hanzo, 2000, 2004*a*,*b*).

7.1.1 Summary

In solving an optimisation problem using GA, the parameter set of optimisation is to be coded as a finite length of string. A fitness function is used to evaluate the fitness of different strings. Three operators, namely reproduction, crossover and mutation, are used in a GA. The simulation is started from a random initial population and a new population is generated by applying the three operators stated earlier. The newly generated population is then used to generate the next population. Most of the algorithms stated in this section are based on this approach. The major difference lies only in the representation of different steps mentioned and their implementation. We give a general algorithm of how the CAP is solved using GA (Ghosh et al., 2003).

7.2 Evolutionary strategy

The evolutionary strategy (ES) method was first introduced by Rechenberg (1973). It is based on an encoded representation of the solutions. Each candidate solution is associated with an objective value. The objective value is representative of the candidate solution's performance in relation to the parameter being optimised. It also gives an indication of a candidate solution's performance with respect to other potential solutions in the space. Based on the fitness values, a number of individuals is selected and genetic operators (mutation and/or recombination) are applied to generate new individuals in the next generation. The best solutions generated in one generation becomes the parents for the next generation. Evolutionary strategy is an iterative method and the process of selection and application of genetic operators is repeated until some terminating criterion is reached. When the termination criterion is reached, the best solution to the problem is represented by the best individual in all generations. The basic steps of the evolutionary strategy method are summarised in algorithm 7.2.

The ES algorithm has the advantage of producing reliable solutions in a smaller number of generations as compared to other heuristics such as the genetic algorithm. This is because at each generation only one parent produces all the feasible solutions.

Sandalidis et al. (1998) introduced the combinatorial evolution strategy (CES) which belongs to the class of evolutionary algorithm (EAs). CES was first introduced in Nissen (1993, 1994) and was shown to be a fast method of solving quadratic assignment problems. The CES was applied to the DCA, HCA and the BCA where each one was formulated as an optimisation problem. Both uniform and non-uniform traffic conditions were considered. The authors claimed that best results were obtained for the DCA case.

Algorithm 7.2: Evolutionary strategy algorithm

1 - Generate an initial population of λ individuals
2 - Evaluate each individual according to a fitness function
3 - Select μ best individuals and discard the remaining ones
4 - Apply genetic operator to create λ offsprings from μ parents
5 - Go to step 2 until a desired solution has been found or predetermined number of generations has been produced and evaluated

Vidyarthi et al. (2005) presented ES which optimised the channel assignment. The ES approach used an efficient problem representation as well as an appropriate fitness function. A novel hybrid channel-assignment-based scheme called D-ring had been presented. The D-ring method yielded a faster running time and simpler objective function. Furthermore, it proposed a novel way of generating the initial population which reduced the number of channel reassignments and therefore yielded a faster running time and may generate a better initial parent. Vidyarthi et al. (2005) have mainly compared their new evolutionary strategy with the one proposed in Sandalidis et al. (1998). The authors also proposed a new HCA strategy based on the fixed reuse distance concept. Each base station had a controller; the status of all calls and changes in each cell were being sent to all the other cells using a good wired network between the computers of all cells. Channel assignment was made by the controller of the concerned base station according to the knowledge about the neighbours of a given cell. In this paper, an ES-based approach is presented with an efficient problem representation and a simplified fitness function. The fitness function took care of the soft constraints whereas the hard constraints were taken care of by the problem representation and the proposed new allocation scheme. The proposed algorithm used integers to represent the solution vector. The time complexity of the proposed algorithm in Vidyarthi et al. (2005) is far less than the one proposed in Sandalidis et al. (1998).

Creput et al. (2005) looked at another dimension of the CAP problem. The main focus was on the dimensioning process of cellular networks which in turn would have a direct impact on the cost on the equipment used to cover an area such as a city for example. The authors have designed an irregular hexagonal cells in an adaptive way subject to the traffic density and geometrical constraints. This process is known as adaptive meshing. The mesh generation problem for mobile network used an evolutionary algorithm known as hybrid island evolutionary strategy (HIES). Simulations were performed in real-life situations and have shown it to be a promising algorithm.

7.2.1 Summary

The ES-based algorithm produces reliable solutions in a small number of generations compared to other heuristics. This is because at each generation only one parent produces all the feasible solutions. The proposed algorithm by Vidyarthi et al. (2005) uses integers to represent solution vectors and thus reduces the computational time of the one presented by Sandalidis et al. (1998). Furthermore, the neighbouring area concept, introduced in this paper, has the advantage of not selecting channels that would result in co-channel interference.

Bibliography

Beckmann, D. & Killat, U. (1999). A new strategy for the application of genetic algorithms to the channel-assignment problem. *IEEE Transactions on Vehicular Technology* 48(4), 1261–1269

Bremermann, H. (1958). The evolution of intelligence: The nervous system as a model of its environment. Technical Report 1. University of Washington, Seattle. Department of Mathematics.

Creput, J., Koukam, A., Lissajoux, T. & Caminada, A. (2005). Automatic mesh generation for mobile network dimension using evolutionary. *IEEE Transactions on Evolutionary Computation* 9(1), 18–30

Cuppini, M. (1994). A genetic algorithm for the channel assignment problems. *European Transactions on Telecommunications* 5(2), 285–294

Ergun, C. & Hacioglu, K. (2000). Multiuser detection using a genetic algorithm in CDMA communications systems. *IEEE Transactions on Communications* 48(8), 1374–1383

Fraser, A. (1957). Simulation of genetic systems by automatic digital computers. II, Effects of linkage on rates under selection. *Australian J. of Biol Sci* 10, 492–499

Fu, X., Pan, Y. & Bourgeois, A. N. (2003). A three-stage heuristic combined genetic algorithm strategy to the channel assignment problem. In Proceedings of the International Parallel and Distributed Processing Symposium

Funabiki, N. (2000). A three-stage heuristic combined neural network algorithm for channel assignment in cellular mobile systems. *IEEE Transactions on Vehicular Technology* 49(2), 397–403

Gamst, A. (1986). Some lower bounds for a class of frequency assignment problems. *IEEE Transactions on Vehicular Technology* VT- 35(1), 8 – 14

Ghosh, S., Sinha, B. & Das, N. (2003). Channel assignment using genetic algorithm based on geometry symmetry. *IEEE Transactions on Vehicular Technology* 52(4), 860–875

Ghosh, S., Sinha, B. & Das, N. Optimal channel assignment in cellular networks with non-homogeneous demands. *in* 'Proceedings of the 56th IEEE Vehicular Technology Conference'. Vol. 3, Vancouver, Canada. pp. 1736–1743

Goldberg, D. (1989). *Genetic Algorithms in Search, Optimisation, and Machine Learning.* Addisson-Wesley. Reading, Massachusetts, USA

Holland, J. (1975). *Adaptation in Natural and Artificial Systems.* The University of Michigan Press. Michigan

Juntti, M.J. (1997). Genetic algorithms for multiuser detection in synchronous CDMA. 'Proceedings IEEE International symposium on Information Theory'. pp. 492–492

Kim, J., Park, S., Dowd, P. & Nasrabadi, N. (1995). Genetic algorithms approach to the channel assignment problem. 'Proceedings of the Asia Pacific Conference on Communications'. pp. 564–567

Kim, J., Park, S., Dowd, P. & Nasrabadi, N. (1996). Channel assignment in cellular radio using genetic algorithms. *Wireless Personal Communications* 3, 273–286

Lai, W. (1996). Channel assignment through evolutionary optimisation. *IEEE Transactions on Vehicular Technology* 45(1), 91–96

Lim, K., Kumar, M. & Das, S. (1999). Message ring based channel reallocation scheme for cellular networks. 'Proceedings of the International Symposium on Parallel Architectures, Algorithms, and Networks'. pp. 426–431

Lima, M., Araujo, A. & Cesar, A. (2002). Dynamic channel assignment in mobile communications based on genetic algorithms. *in* 'Proceedings of the 13 th IEEE International Symposium on Personnal, Indoor and Mobile Radio Communication, 15-18 September'. pp. 2204–2208

Ngo, C. Y. & Li, V. (1998). Fixed channel assignment in cellular radio networks using a modified genetic algorithm. *IEEE Transactions on Vehicular Technology* 47(1), 163–172

Nissen, V. (1993). A new efficient evolutionary algorithm for the quadratic assignment problem. Springer- Verlag Heildelberg pp. 279–267

Nissen, V. (1994). Solving the quadratic assignment problem with clues from nature. *IEEE Transactions on Neural Networks* 5(1), 66–72

Rechenberg, I. (1973). Evolutions strategies: Optimierung technischer systeme nach prinzipien der biologischen evolution. *Frommann-Holzboog, Stuttgart*

Revuelta, L. (2007). A new adaptive genetic algorithm for fixed channel assignment. *Information Sciences* 177(13), 2655–2678

Sandalidis, H., Strvroulakis, P. & Rodriguez-Tellez, J. (1998). An efficient evolutionary algorithm for channel resource management in cellular mobile systems. *IEEE Transactions on Vehicular Technology* 2(4), 125–137

Santos, V., Conceio, M., Pereira, V., Dinis, M. & Neves, J. (2000). A new dynamic channel allocation technique: Simplified maximum packing. 'Proceedings of the IEEE 52nd Vehicular Technology Conference'. Vol. 2. pp. 1390–1394

Santos, V., Dinis, M. & Neves, J. (2001). Inclusion of optimisation methods on a new dynamic channel allocation scheme. *in* F. d Foz, ed., 'Confetele'. Portugal. pp. 581–585

Sung, C. and Wong, W. (1997). Sequential packing algorithm for channel assignment under co-channel and adjacent channel interference constraint. *IEEE Transactions on Vehicular Technology* 46(3), 676–686

Tcha, D., Chung, Y. & Choi, T. (1997). A new lower bound for the frequency assignment problem. *IEEE/ACM Transactions on Networking* 5(1), 34–39

Vidyarthi, G., Ngom, A. & Stojmenovic, I. (2005). A hybrid channel assignment approach using an efficient evolutionary strategy in wireless mobile networks. *IEEE Transactions on Vehicular Technology* 54(5), 1887–1895

Wong, S. (2003). Channel allocation for broadband fixed wireless access networks. Unpublished PhD thesis. University of Cambridge. UK

Wong, S. & Wassell, I. (2002*a*). Application of game theory for distributed dynamic channel allocation. 'Proceedings of the 55 th IEEE Vehicular Technology Conference'. Vol. 1. pp. 404–408

Wong, S. & Wassell, I. (2002*b*). Dynamic channel allocation using a genetic algorithm for a TDD broadband fixed wireless access network. 'Proceedings of the IASTED International Conference on Wireless and Optical Communications'. pp. 521–526

Yen, K. & Hanzo, L. (2000). Hybrid genetic algorithm based multiuser detection schemes for synchronous CDMA systems. 'Proceedings of the 51 st IEEE Vehicular Technology Conference'. pp. 1400–1404

Yen, K. & Hanzo, L. (2004*a*). Genetic algorithm assisted joint multiuser symbol detection and fading channel estimation for synchronous CDMA systems. *IEEE Transactions on Vehicular Technology* 53(5), 1413–1422

Yen, K. & Hanzo, L. (2004*b*). Genetic algorithm assisted multiuser detection in asynchronous CDMA communications. *IEEE Transactions on Vehicular Technology* 53(5), 1413–1422

Yener, A. & Rose, C. (1997). Genetic algorithms applied to cellular call admission problem: Local policies. *IEEE Transactions on Vehicular Technology* 46(1), 72–79

Zomaya, A. Y. (2002). Observations on using genetic algorithms for channel allocation in mobile computing. *IEEE Transactions on Parallel and Distributed Systems* 13(9), 948–962

Chapter 8

Q-learning and other algorithms

8.1 Q-learning approach

Another approach of solving the dynamic channel allocation problem was Q-learning-based (Sutton and Barto, 1998) proposed by (Nie and Haykin, 1999*b*,*a*). The main aim of the Q-learning approach was to look at the DCA problem through a form of real-time reinforcement learning (Barto et al., 1995). The system was designed to learn an optimal assignment policy by directly interacting with the mobile communication environment, the environment with which it works. Learning is accomplished progressively by appropriately utilising the past experience which is obtained during real-time operation. The performance of Q-Learning was compared mainly with FCA and MAX-AVIAL (Sivaranjan et al., 1990). Later, El-Alfy et al. (2001) proposed an algorithm that prioritised handoff call requests over new call requests. The problem was formulated as an average cost continuous time Markov decision problem. Then a model-based reinforcement learning approach was developed for finding a self-adjusting allocation strategy that is approximately optimal. The algorithm was tested only for a single traffic class. The Q-learning algorithm is described as follows: let the environment with which a learner interacts be a finite state discrete time stochastic dynamical system. X is defined as the set of possible states, $X = \{x_1,\ x_2, ..., x_n\}$ and A is defined as a set of possible actions, $A = \{a_1,\ a_2, ..., a_n\}$. The interaction between the learner and the environment at each time is given in Algorithm 8.1.

Algorithm 8.1: Q-learning algorithm

---

1 - The learner senses the state $x_t \in X$
2 - Based on x_t, the learner chooses an action $a_t \in A$ to perform
3 - As a result, the environment makes a transition to the
new state $x_{t+1} = y \in X$ according to probability $P_{xy}(a_t)$,
and thereby generates a return cost r_t
4 - The r_t is passed back to the learner and the process is repeated

---

The objective of the learner is then to find an optimal policy $\pi^*(x) \in A$ for each x, which minimises some cumulative measure of the costs $r_t = r(x_t, a)$ received over time.

8.1.1 Summary

A self-learning scheme based on Q-learning was studied. The approach was applied to DCA in a 49-cell mobile communication and the results were promising. However, if the approach is to be implemented in a real system, more studies need to be done.

8.2 Other algorithms

From literature it can be found that some authors have used different techniques. A pattern approach which fits to the CAP was used in Kim and Kim (1994) with the assumption that there is no channel interference between two cells separated by more than a certain distance. Based on this approach, the CAP was being formulated into a manageable optimisation problem and a two-phase heuristic algorithm was being devised to find the solution. Leese (1997) had considered the CAP on a lattice of hexagonal cells with allowed assignments generated by regular tiling of a single polyhex. The main aim was to investigate the interplay between the co-channel and the adjacent channel separations without restricting the assignments unnecessarily. Tcha et al. (2000) had presented a perturbation minimising frequency assignment method. The main aim of the method was to assign available frequencies for newly generated requirements with the minimum change in the existing frequency assignments while still meeting the interference-related constraints. Gandhi et al. (2003) have looked at the channel allocation problem in broadcast networks using approximation algorithms. The aim was to minimise the maximum load on any channel. Fernando and Fapojuwo (2002) presented a new algorithm known as the Viterbi-Like Algorithm (VLA), which is very similar to the sequential trellis search algorithm of the original Viterbi algorithm used in the field of information decoding (Jr, 1973; Viterbi, 1967). The VLA minimised bandwidth by removing redundant channel assignments from further consideration and keeping only the survivors after each step of the channel assignment. It thus reduced the number of iterations to be performed in the subsequent steps of channel assignment, hence achieving a faster execution time. Ghosh et al. (VTC Fall 2002, 2003) did more research work in the field of CAP. In Ghosh et al. (2003) the concept of partitioning the critical block into several smaller sub-networks was introduced. This idea of partitioning was then extended for assigning frequencies to the rest of the network. The proposed algorithm gave optimal results in terms of bandwidth requirements and computation time. Some years later, Ghosh et al. (2006) introduced a technique known as coalesced CAP, which reduced the computational time drastically. The aim in the paper presented by Ghosh et al. (2006), was to reduce the search space and apply appropriate algorithm to assign frequencies and then use a modified version of forced assignment with rearrangement (FAR) (Tcha et al., 2000) to produce an optimal solution for the whole space. Recently, Kim et al. (2007) presented a new encoding for the minimum span frequency assignment problem. The new algorithm reduced the search space dramatically and a combination of a genetic algorithm global search and a computationally efficient local search method was used. This method is known as mimetic algorithm and was applied to the Philadelphia problem to compare with previous results.

Bibliography

Barto, A., Bradtke, S. & Singh, S. (1995). Learning to act using real-time dynamic programming. *Artificial Intelligence* 72 (1-2), 81–138

El-Alfy, E., Yao, Y.-D. & Heffes, H. (2001). A model based q-learning scheme for wireless channel allocation with prioritized handoff. 'Proceedings of the Global Telecommunications Conference GLOBECOM'01.IEEE'. Vol. 6. pp. 3668–3672

Fernando, X. & Fapojuwo, A. (2002). A viterbi-like algorithm with adaptive clustering for channel assignment in cellular radio networks. *IEEE Transactions on Vehicular Technology* 51(3), 73–87

Forney Jr, G. F. (1973). The Viterbi algorithm. *IEEE* 61(3), 268–278

Gandhi, R., Khuller, S., Srinivasan, A. & Wang, N. (2003). Approximation algorithms for channel allocation problems in broadcast networks. *Lecture Notes in Computer Science* 2764, 47–58

Ghosh, S., Sinha, B. & Das, N. (VTC Fall 2002). Optimal channel assignment in cellular networks with non-homogeneous demands. Proceedings of the 56 th IEEE Vehicular Technology Conference. Vol. 3. Vancouver, British Columbia, Canada. pp. 1736–1743

Ghosh, S., Sinha, B. & Das, N. (2003). A new approach to efficient channel assignment for hexagonal cellular networks. *International Journal of Foundations of Computer Science* 14(3), 439–463

Ghosh, S., Sinha, B. & Das, N. (2006). Coalesced CAP: An improved technique for frequency assignment in cellular networks. *IEEE Transactions on Vehicular Technology* 55(2), 640–653

Kim, S. & Kim, S. (1994). A two phase algorithm for frequency assignment in cellular mobile systems. *IEEE Transactions on Vehicular Technology* 43(3), 542–548

Kim, S., Smith, A. & Lee, J. (2007). A memetic algorithm for channel assignment in wireless FDMA systems. *Computers and Operations Research* 34(6), 1842–1856

Leese, R. A. (1997). A unified approach to the assignment of radio channels on a regular hexagonal grid. *IEEE Transactions on Vehicular Technology* 46(4), 968–980

Nie, J. & Haykin, S. (1999*a*). A dynamic channel assignment policy through Q-learning. *IEEE Transactions on Neural Networks* 10(6), 1443–1455

Nie, J. & Haykin, S. (1999*b*). A Q-learning-based dynamic channel assignment technique for mobile communication systems. *IEEE Transactions on Vehicular Technology* 48(5), 1676–1687

Sivaranjan, K., McEliece, R. & Ketchum, J. (1990). Dynamic channel assignment in cellular radio. *Procedings of the IEEE 40 th Vehicular Technology Conference*, pp. 631–637

Sutton, R. S. & Barto, A. G. (1998). *Introduction to Reinforcement Learning*, Cambridge, USA: MIT

Tcha, D., Kwon, J., Choi, T. & Oh, S. (2000). Perturbation minimising frequency assignment in a changing TDMA/FDMA cellular environment. *IEEE Transactions on Vehicular Technology* 49(2), 390–396

Viterbi, A. J. (1967). Error bounds for convolutional codes and an asymptotically optimum decoding algorithm. *IEEE Transactions Information Theory* IT-13, 260–269

Chapter 9

The channel allocation problem from a multi-objective view

9.1 Multi-objective approach

In this chapter, we present the CAP from a multi-objective perspective, which is a new era of research in this field. The demand of mobile users is on an increasing trend as more and more people make use of mobile communications in everyday life. With the new approach that is being proposed, it will be easier to manage the network system in case there are changes in demand. The network manager will have several options for deciding which solution best fits a particular situation. The multi-objective approach gives several possible solutions in terms of blocked calls and redundant frequency from the Pareto set.

While solving the CAP using a single objective, we find that one obtains blocked calls; on the other hand, lots of frequencies remain unused from the frequency bandwidth. Most of the researchers have tried to obtain the optimal solution but if it is not attained, blocked calls result. Ideally, one should aim to minimise these blocked calls while solving the CAP. As mentioned earlier, in the frequency bandwidth there are frequencies that are unused which we have called redundant frequencies. These could be used in cases when demand is increasing but obviously satisfying the interference constraints. Thus, one should aim to maximise the use of redundant frequencies. Using the blocked calls and redundant frequencies, we define the multi-objective dimensionality of the CAP. Therefore, in this chapter we describe how the tabu search is applied to the CAP and how the blocked calls and redundant frequencies are calculated. Our aim is to minimise the blocked calls and to maximise the redundant frequencies. We then use the NSGA II algorithm to optimise these two objective functions. The results obtained have been published in the proceedings of the IEEE World Congress on Computational Intelligence 2008, pp. 3914–3917 in the paper titled 'A multi-objective approach to the channel assignment problem'.

9.2 Experimental results

We describe the performance of different methods (tabu search and NSGA II) for solving the channel assignment problem. The ordering of the cells is a key factor for efficient frequency assignment (Beckmann and Killat, 1999) and Ghosh et al. (2006) and a tabu search algorithm is employed to learn the order. The algorithm we have used is simple and easily applicable.

For each problem, we choose three values M_1, M_2 and M_3 for the maximum frequency bandwidth: M_1 is a value less than the theoretical lower bound presented by Gamst (1986), M_2 is equal to the theoretical lower bound and M_3 is greater than the theoretical lower bound.

Starting with a random permutation of the cells, the intensification phase using the $firstN2$ technique is applied, followed by the diversification phase. These phases are applied for some predefined number of iterations or the procedure stops when a certain number of global iterations are performed without any improvement in the move values. One global iteration consists of the intensification phase followed by the application of the diversification phase. At each iteration, the algorithm learns the order of the cells, which minimises the blocked calls and satisfies the interference constraints. At the end of each iteration, the redundant frequency is calculated. For each parameter value of $M_{i=1,2,3}$, 200 global iterations are performed and a plot of blocked calls against redundant frequencies is carried out. The method is tested on the commonly used benchmark Philadelphia problem, which has a 20-cell system. The cellular layout of the considered 21-cell layout is depicted in Figure 3.1.

The two methods described in earlier chapters have been applied to the well-known benchmark problems. It is well known from literature that for problems 2 and 6 (from Table 3.7 in chapter 3), admissible frequencies within the theoretical lower bound and a reasonable time are difficult to obtain. For each set of problems, the number of blocked calls and the redundant frequencies are calculated using both methods. Having obtained these data, the aim was to minimise the blocked calls and maximise the redundant frequencies. The multi-objective algorithm, NSGA-II, is used to obtain the Pareto front and is applied a few times to each of the data sets. The plots of the different cases obtained are shown in Figure 9.1. The Pareto front is denoted by the circles in each plot. The front denotes a trade-off between the blocked calls and the redundant frequencies. These are helpful in case of a sudden rise in demand and the optimal solution should lie in the Pareto front for better decision-making. It is observed that for problems for which are easy to obtain the optimal solution, the Pareto front lies on the zero blocked axis. But for the difficult cases – that is, cases 2 and 6 – the Pareto front is clear and distinct, in other words we obtain better results for these cases with the multi-objective approach.

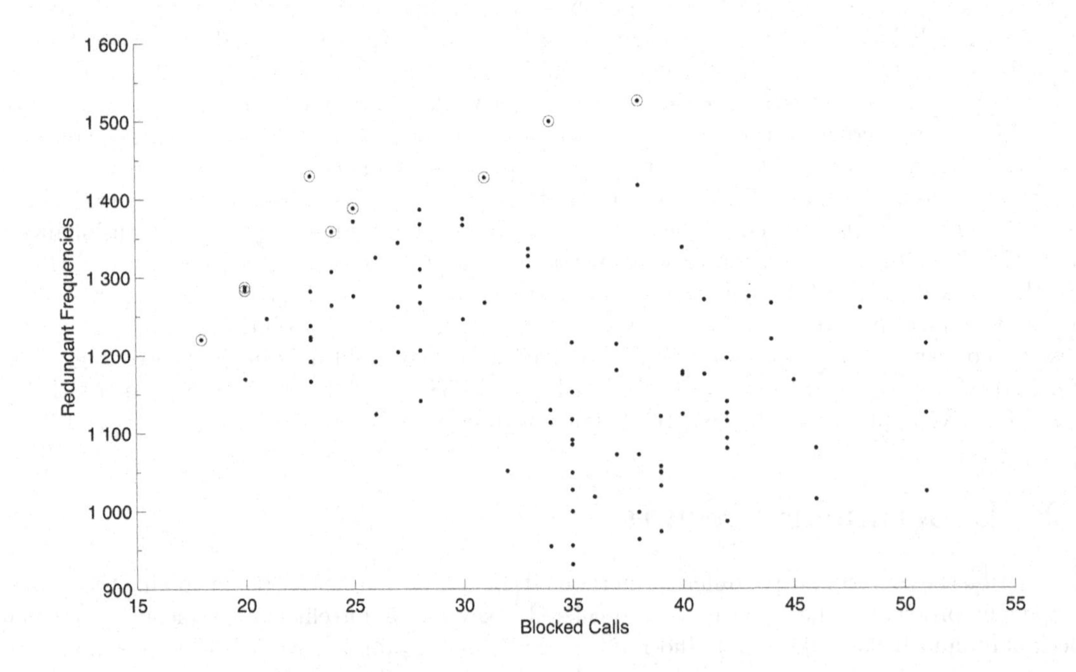

Figure 9.1: Pareto front for problem 2 (Table 3.7)

9.3 Comparison of different approaches

In this section we shall compare the performances of the different approaches based on the eight Philadelphia benchmark problems mentioned earlier in the Chapter 3. Some authors have used the Kunz's benchmark, but for the comparison in this section we are focusing only on the Philadelphia problems. Among the eight benchmarks, other than problems 2 and 6, the optimal solution for the remaining six are easily obtained. The reason is that, for these six cases, the required bandwidth is mainly limited by the co-channel interference constraint only. The most challenging cases are problems 2 and 6. For example, the lower bound for problem 2 is 253 channels, and in Ngo and Li (1998) 165 hours were required to solve the problem on an HP Apollo 9000/700 workstation with 268 channels. A comparison of the different approaches is given in Table 9.1.

Table 9.1: Comparison of results for the different approaches to solve the CAP

Problems	1	2	3	4	5	6	7	8
Lower Bounds	381	427	533	533	221	253	309	309
Vieira et al. (2008)	381	427	533	533	221	253	309	309
Fu et al. (2006)	381	432	533	533	-	253	-	309
Ghosh et al. (2006)	381	427	533	533	221	253	309	309
Kendall and Mohamad (2004)	381	533	533	-	265	-	309	
Ghosh et al. (2003)	381	427	533	533	221	253	309	309
Fernando and Fapojuwo (2002)	381	438	533	533	-	266	-	309
Chakraborty (2001)	381	463	533	533	221	273	309	309
Battiti et al. (2001)	381	427	533	533	221	254	309	309
Tcha et al. (2000)	381	433	533	533	-	260	-	309
Beckmann and Killat (1999)	381	427	533	533	221	253	309	309
Ngo and Li (1998)	-	-	-	-	221	268	-	309
Kim et al. (1997)	381	-	533	533	221	-	309	309
Sung and Wong (1997)	381	436	533	533	-	268	-	309
Wang & Rushforth (1997)	381	433	533	533	221	263	309	309
Kim et al. (1996)	381	-	533	533	-	-	-	-
Ko (1994)	381	464	533	536	-	293	-	310
Funabiki and Takefuji (1992)	381	-	533	533	221	-	309	309
Sivaranjan et al. (1989)	381	447	533	533	-	270	-	310

A comparison of the results obtained from the different approaches is also included. It is also interesting to note that it is most challenging to obtain optimal solutions for problems 2 and 6, as can be deduced from Table 9.1, and this can be explained by the fact that CAP is dependent on the ordering of the cells.

9.4 Summary

This chapter presented the novelty of the work that is being done where a multi-objective dimension to the CAP is introduced. The methods described in Chapters 4 and 5 have been applied to the CAP - that is, we first apply the tabu search algorithm to calculate the blocked calls and redundant frequencies and then, using these two objectives, NSGA II is applied to obtain the Pareto front. In the tabu search algorithm we have defined a weight function which will bias the search to those cells that have higher demand. Computational time is an important factor that researchers consider in solving the CAP, especially in practical situations. At the end of this chapter we also gave a summary of the solutions to the CAP problem, namely the Philadelphia problem, discussed in the previous chapters. This gives an idea of which methods achieve the theoretical lower bound.

Bibliography

Battiti, R., Bertossi, A. & Cavallaro, D. (2001). A randomized saturation degree heuristic for channel assignment in cellular networks. *IEEE Transactions on Vehicular Technology* 50(2), 364–374

Beckmann, D. & Killat, U. (1999). A new strategy for the application of genetic algorithms to the channel-assignment problem. *IEEE Transactions on Vehicular Technology* 48(4), 1261–1269

Chakraborty, G. (2001). An efficient heuristic algorithm for channel assignment problem in cellular radio networks. *IEEE Transactions on Vehicular Technology* 50(6), 1528–1539

Fernando, X. & Fapojuwo, A. (2002). A viterbi-like algorithm with adaptive clustering for channel assignment in cellular radio networks. *IEEE Transactions on Vehicular Technology* 51(3), 73 –87

Fu, X., Bourgeois, A., Fan, P. & Pan, Y. (2006). Using genetic algorithm approach to solve the dynamic channel-assignment problem. *International Journal of Mobile Communications* 4(3), 333–353

Funabiki, N. & Takefuji, Y. (1992). A neural network parallel algorithm for channel assignment problems in cellular radio networks. *IEEE Transactions on Vehicular Technology* 41(4), 430–437

Gamst, A. (1986). Some lower bounds for a class of frequency assignment problems. *IEEE Transactions on Vehicular Technology* VT- 35(1), 8–14

Ghosh, S., Sinha, B. & Das, N. (2003). Channel assignment using genetic algorithm based on geometry symmetry. *IEEE Transactions on Vehicular Technology* 52(4), 860–875

Ghosh, S., Sinha, B. & Das, N. (2006). Coalesced CAP: An improved technique for frequency assignment in cellular networks. *IEEE Transactions on Vehicular Technology* 55(2), 640–653

Kendall, G. & Mohamad, M. (2004). Channel assignment in cellular communications using a great deluge hyper-heuristic. *in* 'Proceedings of the 12 th IEEE International Conference on Networks'. Vol. 2. pp. 769–773

Kim, J., Park, S., Dowd, P. & Nasrabadi, N. (1996). Channel assignment in cellular radio using genetic algorithms. *Wireless Personal Communications* 3, 273–286

Kim, J., Park, S., Dowd, P. & Nasrabadi, N. (1997). Cellular radio channel assignment using a modified hopfield network. *IEEE Transactions on Vehicular Technology* 46(4), 957–967

Ko, T. (1994). A frequency selective insertion strategy for fixed channel assignment. *in* 'Proceedings of the 5th IEEE International Symposium Personnal, Indoor and Mobile Radio Communications'. pp. 311–314

Ngo, C. Y. & Li, V. (1998). Fixed channel assignment in cellular radio networks using a modified genetic algorithm. *IEEE Transactions on Vehicular Technology* 47(1), 163–172

Sivaranjan, K., McEliece, R. & Ketchum, J. (1989). Channel assignment in cellular radio. Vol. VTC 89. pp. 846–850

Sung, C. & Wong, W. (1997). Sequential packing algorithm for channel assignment under co-channel and adjacent channel interference constraint. *IEEE Transactions on Vehicular Technology* 46(3), 676–686

Tcha, D., Kwon, J., Choi, T. & Oh, S. (2000). Perturbation minimising frequency assignment in a changing TDMA/FDMA cellular environment. *IEEE Transactions on Vehicular Technology* 49(2), 390–396

Vieira, C. E., Gondim, P. R. L., Rodrigues, C. & Bordim, J. (2008). A new technique to the channel assignment problem in mobile communication networks. *in* 'Proceedings of the IEEE 19th International Symposium on Personal Indoor and Mobile Commincations'. pp. 1–5

Wang, W. & Rushforth, C. (1997). Local search for channel assignment in cellular mobile networks. *DIMACS Series in Discrete Mathematics and Theoretical Computer Science* 35, 689–709